Explaining Games

SYNTHESE LIBRARY

STUDIES IN EPISTEMOLOGY, LOGIC, METHODOLOGY, AND PHILOSOPHY OF SCIENCE

VOLUME 346

For further volumes:
http://www.springer.com/series/6607

Explaining Games

The Epistemic Programme in Game Theory

by

Boudewijn de Bruin
University of Groningen, The Netherlands

Springer

Boudewijn de Bruin
Faculty of Philosophy
University of Groningen
Oude Boteringestraat 52
9712 GL Groningen
The Netherlands
b.p.de.bruin@rug.nl

ISBN 978-94-007-3331-2 ISBN 978-1-4020-9906-9 (eBook)
DOI 10.1007/978-1-4020-9906-9
Springer Dordrecht Heidelberg London New York

Softcover reprint of the hardcover 1st edition 2010

Printed on acid-free paper

Springer is part of Springer Science+Business Media (www.springer.com)

When you collect marine animals there are certain flat worms so delicate that they are almost impossible to capture whole, for they break and tatter under the touch. You must let them ooze and crawl of their own will onto a knife blade and then lift them gently into your bottle of sea water.

—John Steinbeck, *Cannery Row*

for my parents and my teachers

Contents

Acknowledgements

Many people have helped me in one or more of the stages that have led to this book. Johan van Benthem and Martin Stokhof supervised the doctoral research with which it all started, and I am immensely grateful for their inspiration and their care. Peter van Emde Boas, Govert den Hartogh, Theo Kuipers, Wlodek Rabinowicz, Robert Stalnaker, Frank Veltman and Gerard de Vries probed my ideas as members of the dissertation committee in constructive ways which, I hope, have not failed in their effect. I also owe special thanks to Mark van Atten, Christina Bicchieri, Giacomo Bonanno, Adam Brandenburger, Mikaël Cozic, Catarina Dutilh-Novaes, Paul Egré, Bob Galesloot, Joseph Halpern, Paul Harrenstein, Martin van Hees, Frank Hindriks, Wiebe van der Hoek, Barteld Kooi, Allessandro Lanteri, Rosja Mastop, Philippe Mongin, Rohit Parikh, Marc Pauly, Herman Philipse, Ian Priestnall, Philip Reny, the late Michael Robins, Menno Rol, Jan-Willem Romeijn, Robert van Rooij, Hans Rott, Ariel Rubinstein, Gabriel Sandu, Allard Tamminga, Bruno Verbeek and Bauke Visser, the anonymous referees of the journals in which parts of the present book have appeared earlier, audiences in Amsterdam, Dagstuhl, Groningen, Lund, Nunspeet, Paris, Prague, Rotterdam and Turin, and my editor at Springer, Ingrid van Laarhoven. I would like to thank the Netherlands Organisation for Scientific Research (NWO) and the De Bussy Foundation for their generous financial support of research projects related to this book. Finally, I am honoured to be able to dedicate this book to my parents and my teachers.

Part of Chapter 2 appeared as 'Common Knowledge of Payoff Uncertainty in Games', *Synthese*, 163(1) July (2008), 79–97; ©2007 Boudewijn de Bruin

Part of Chapter 3 appeared as 'Common Knowledge of Rationality in Extensive Games,' *Notre Dame Journal of Formal Logic*, 49(3) (2008), 261–280; ©2008 University of Notre Dame. Reproduced by permission of Duke University Press

Part of Chapter 4 appeared as 'Reducible and Nonsensical Uses of Game Theory,' *Philosophy of the Social Sciences*, 38(2) June (2008): 247–266; ©2008 Sage Publications. Reproduced by permission

Part of Chapter 4 appeared as 'On the Narrow Epistemology of Game Theoretic Agents' in Ondrej Majer, Ahti-Veikko Pietarinen and Tero Tulenheimo (eds.),

Games: Unifying Logic, Language, and Philosophy, series *Logic, Epistemology and the Unity of Science* (Berlin: Springer, 2009) 27–36; ©2009 Springer Science + Business Media B. V. Reproduced by permission

Part of Chapter 5 appeared as 'Overmathematisation in Game Theory: Pitting the Nash Equilibrium Refinement Programme against the Epistemic Programme,' *Studies in History and Philosophy of Science Part A*, 40(3) (2009), 290–300; ©2009 Elsevier Ltd. Reproduced by permission

Introduction

Game theorists may appreciate Schiller's famous observation that

> Human beings are only fully human when they play games.

This suggests that game theory is a mathematical theory of almost all forms of human agency. But is Schiller right? The aim of this book is to find out.

I will undertake two projects. The first, reported in Part I, studies game theory internally. I will develop a formalism in epistemic logic to expose assumptions that game-theoretic solution concepts make about human agency. While the core of this research is interpretative and philosophical, and while the exact set-up of the logic is inspired by philosophical, rather than game-theoretic questions, a number of new logical results are proven along the way. In this way, the book not only contributes to the philosophy of game theory, but also to the Epistemic Programme within game theory—a branch of the social sciences, also known as Interactive Epistemology, in which logicians, probability theorists, economists and game theorists combine efforts to increase our understanding of human strategic interaction.

The second project, covered in Part II, studies non-cooperative game theory from an external, epistemological perspective. I will argue that game theory does not make sense as a normative theory, and that it can be reduced to decision theory whenever it aims to explain actual human behaviour. Where the first project undertakes a careful reading of the internal workings of the mathematical models proposed in non-cooperative game theory, the second project focuses on the methodological assumptions of the researchers, and on their research strategies. A case study concludes the external investigations, contrasting the Nash Equilibrium Refinement Programme and the Epistemic Programme.

Chapter 1 introduces epistemic characterisation theorems and presents the logical form of game-theoretic modelling in normal form and extensive games. Characterisation theorems not only lie at the core of the Epistemic Programme in game theory, but are also essential to the argument about reducing game theory to decision theory and in the investigations of the Nash Equilibrium Refinement Programme. I will argue that these theorems are best understood as involving conditions on the possible actions that players can choose between, the preferences over the possible outcomes of the combined play of all those possible actions, the rationality principles that the players decide to act upon, the actions that the players actually perform in the game-playing situation, and the beliefs of the players about all five ingredients—that is, with intended recursion, the possible actions, preferences, principles, actions to be performed, and beliefs. I will examine the exact content of these conditions for normal form and extensive games, introducing two interpretations for the latter kind of games. In the one-shot interpretation, playing an extensive game is playing the normal form of that game, and all players pick one full strategy simultaneously. Epistemic characterisations for the one-shot interpretation follow the logical form of epistemic characterisations for normal form game solution concepts. The many-moment interpretation, by contrast, stipulates that playing an extensive game involves more than one decision moment, and that at each decision moment precisely one player chooses an individual action that leads to a future decision moment, or to the termination of the game. The rest of the chapter sets out the logical preliminaries required in the remainder of the book. As well as recalling some standard ingredients of epistemic logic, Chapter 1 provides a brief, systematic impression of the formalisation presented in Part I.

Chapter 2 presents a logical formalism in order to describe epistemic characterisation results of normal form game solution concepts in a uniform way which enables detailed philosophical discussion of their internal structure. Two epistemic characterisation theorems of the Nash equilibrium are considered, one involving strategies, the other beliefs. A central claim here is that the former does not conform to the belief–desire framework of human agency unless the beliefs of the players are taken to be necessarily, conceptually true (that is, unless the T-axiom of veridicality is assumed). Another claim is that the meaningfulness of the latter characterisation result, concerning beliefs in equilibrium, depends upon the meaningfulness of the former result. Ultimately, the Nash equilibrium does not provide a justification in the endogenous way—still heavily influential in the Epistemic Programme—promoted by John von Neumann and Oskar Morgenstern, because it is only possible to explain the behaviour of players playing the Nash equilibrium if you are prepared to go beyond the game-playing situation by using statistical or other kinds of data that are not interior to game theory.

I will then develop a uniform logical language to represent the epistemic characterisation results of iterated dominance conceptions, which models rationality (expected utility maximisation) inductively and implicitly rather than by explicit reference to players' utility functions and probabilistic beliefs. This approach allows me to demonstrate that certain intuitions underlying a well-known result by Eddie Dekel and Drew Fudenberg about the behaviour of players with less than full information about their opponents' utility functions do not characterise the so-called Dekel–Fudenberg procedure. Rather, they characterise a different iterative dominance concept that is called here *mixed iterated strict weak dominance*. Yet a direct characterisation of the Dekel–Fudenberg procedure can be obtained in our framework, too, if we use insights from Robert Stalnaker's game models. The epistemic characterisation of the Dekel–Fudenberg procedure that Stalnaker obtains is based on a perfected form of rationality where players not only maximise utility given their current beliefs, but also maximise utility given beliefs that they would adopt if their current beliefs were shown to be false. An inductive, implicit axiomatisation of perfect rationality underscores this result.

Chapter 3 turns to extensive games, and studies the main solution concept of backward induction under the one-shot as well as the many-moment interpretation. A reconstruction of one version of the argument that common knowledge would entail backward induction reveals that a one-shot view of extensive game-playing is assumed together with—but strikingly inconsistent with—a rationality principle that takes seriously certain aspects of the game tree structure of extensive games that can only be truly meaningful to the many-moment interpretation. An alternative characterisation can be given that substitutes the incompatible rationality principle for another one; but many game theorists will object to the replacement principle as well. Turning to backward induction, we find that the critique against it needs the many-moment outlook on extensive game-playing along with highly restrictive belief formation practices—which is also incoherent.

Chapter 4 turns to external investigations, first of the rationality principle, and then of game theory used for explanatory and prescriptive, normative purposes. I mention briefly Max Weber's and John Stuart Mill's conceptions of rationality and social explanation, which leads to an investigation of the logical form of rationality as expected utility maximisation and the presuppositions about human agency made by this concept. I will argue that the sharpest, but also the most ambitious phrasing of the principle involves existential quantification only, and that two equally plausible normative interpretations of the principle are obtained by either prefixing the deontic operator *de dicto*, or by infixing it *de re*. Applying the same logic to game-theoretic explanations allows me to defend the claim that if these explanations are to conform to the belief–desire framework—which they should—epistemic characterisation theorems are the canonical purveyors of the players' beliefs. As a result of this, however, game-theoretic explanations become reducible to decision theory. Worse, an epistemological bias embodied in these results towards narrow epistemic policies shows these reductions to be of rather limited applicability. Moreover, with the theory of games being reducible as an explanatory theory, a detailed logical analysis shows that it is nonsensical as a normative theory. All of this should not lead us to reject game theory—non-cooperative game theory, that is—as a whole; but it defers its applications to such domains as mechanism design or conceptual philosophical argumentation in need of equilibrium notions.

Chapter 5 first examines the *true-in-the-abstract* conception of game-theoretic modelling. This conception is seen partly to cause and partly to constitute over-mathematisation of scientific texts, and to lead to an introverted stance to applied mathematics and to a research habit I call *model-tinkering*. The original motivation behind John Nash's solution concept, as well as the ways in which a number of refinement solution concepts are defended in the Nash Equilibrium Refinement Programme betray the true-in-the-abstract view to a considerable extent—in striking contrast to the equally mathematical, but more outward looking Epistemic Programme. I will show that the Nash Equilibrium Refinement Programme and the Epistemic Programme share the same key research objective, that is, to produce the ultimate characterisation of strategic interaction between rational, economic agents. Furthermore, I will argue that the Epistemic Programme has been more successful in reaching this goal, and explain this by reference to its more application-driven mathematical modelling techniques.

Chapter 1
Preliminaries

1.1 The Logic of Game Theory

Human actions can be made sense of in various ways. Combining the dichotomy of understanding and explanation on the one hand, and that of individualism and holism on the other, we can explain actions as expressing meaning, as governed by rules, as fulfilling functions in larger systems, or as being based on individual reasons.[1]

Which of these four modes applies to the theory of games? This question suggests that there is only one way to use game theory in the social sciences, which may not be too plausible given the creativity of the social scientist working in any of the four frameworks and applying game theory in a way she judges productive. Nonetheless, if we take the declared aims of non-cooperative game theory seriously, we are almost automatically led to think of game theory as conforming to the belief–desire framework of action explanation. Game theory explains actions in terms of the reasons agents have to carry them out.[2] While this may not be the only way to make sense of game theory, it is certainly a central sense of game theory. The belief–desire framework forms the basis of this book.

This chapter first deepens our understanding of the contrast between decision theory and game theory, and shows the ways in which to give a precise description of the differences between the two theories in epistemic terms. Subsequently, I will consider normal form and extensive form game-playing in order to distinguish, as their relevant elements, the possible actions a player can choose to perform, her preference ordering and rationality principle, the action eventually performed, and, finally, the beliefs about—with recursion intended—all five ingredients. For extensive games, a one-shot interpretation and a many-moment interpretation will be set apart.

---

[1] Martin Hollis, *The Philosophy of Social Science: An Introduction* (Cambridge: Cambridge University Press, 1994).

[2] The belief–desire framework has most notably been defended by Donald Davidson, *Essays on Actions and Events* (Oxford: Clarendon Press, 1980).

B. de Bruin, *Explaining Games*, Synthese Library 346,
DOI 10.1007/978-1-4020-9906-9_1,

1.1.1 Decision Theory and Game Theory

It is standard to account for the differences between decision and game theory in terms of the number of players, and to note that where decision theory is concerned with one player who has to act in a situation of certainty, uncertainty or risk, game theory is concerned with several players who interact strategically. The number of players, however, is not the crucial factor here. Sometimes, for instance, decision theory is held as the study of individuals playing against a second player, nature. Rather, the difference is that decision theory involves only one agent with beliefs and desires—nature does not have beliefs or desires—and that game-theoretic agents all have beliefs and desires.[3]

This distinctive feature is unequivocally mirrored in the way games against nature and games against opponents with beliefs and desires are represented. The entries of the decision matrix are pairs of real numbers in game theory, but only single real numbers in decision theory. Nonetheless, what these differences lead to—what game-theoretic agents are supposed to do with their beliefs and desires—remains to be examined.

1.1.1.1 The Ban on Exogenous Information

John von Neumann and Oskar Morgenstern describe the function of preferences as follows:

> Every participant can determine the variables which describe his own actions but not those of the others. Nevertheless those 'alien' variables cannot, from his point of view, be described by statistical assumptions. This is because the others are guided, just as he himself, by rational principles—whatever that may mean—and no *modus procedendi* can be correct which does not attempt to understand those principles and the interactions of the conflicting interests of all participants.[4]

That being the case, the difference between decision and game theory is not so much that opponents have beliefs and desires, but rather that any player has to consider her opponents as having such beliefs and desires. The right way to think about your opponent is the way that you think about yourself, as a rational being acting on beliefs and desires, not in the way that you think of the weather or radioactive decay.

One of the most striking consequences of this view—a consequence that recurs throughout the entire book—is that in order to find out what your opponents will do, the only information a game-theoretic agent is supposed to use is her beliefs and desires. Statistical data or any other exogenously based data are out. To conceive of your opponents' agency as guided by rationality means that numerous epistemic

---

[3] In an alternative vocabulary this is the distinction between *parametric* and *strategic* choice situations. Decision theory is also called *rational choice theory*. A classic reference is R. Duncan Luce and Howard Raiffa, *Games and Decisions: Introduction and Critical Survey* (New York: Wiley, 1957).

[4] John von Neumann and Oskar Morgenstern, *Theory of Games and Economic Behavior* (Princeton: Princeton University Press, 1944), 11.

policies that are entirely natural in meteorology or nuclear physics are out. This is quite literally a *ban on exogenous information*. Decision-theoretic models typically have individuals base their probabilistic beliefs on statistics and other outside data. For instance, in order to choose with the help of decision theory whether to build a nuclear plant, statistical data about nuclear catastrophes are essential. Von Neumann and Morgenstern's call to choose one's strategy on the basis of reasoning processes that do not transcend the game structure, but only refer to the rationality principles and the utility functions of the players, excludes such forms of information, however.

In general, various ways to form beliefs about what opponents will play are available. Players may consult statistical surveys, they may call upon experience, they may read historical texts, or they may watch videos of previous games. The ban on exogenous information rules all this out, though, and so the question is: how can game-theoretic agents, if the ban is genuinely enforced, ever obtain information about the prospective strategic choices of their opponents?

To show how merely endogenous information can form the basis of substantial belief formation and lead players to select certain strategies and to exclude others is, in fact, the main research goal of non-cooperative game theory. Up until the 1990s, game-theorists thought that solution concepts that refine the Nash equilibrium would accomplish this goal; but nowadays the Epistemic Programmeis considered the most likely source of insight, particularly by way of mathematical results called *epistemic characterisation theorems*.[5] Strictly obeying the ban on exogenous information, the main idea is that I can derive sensible and precise predictions of my opponents' prospective strategic behaviour from the assumption that they are as rational as I am.[6]

1.1.1.2 Epistemic Characterisation Theorems

In order to get some feel for the general form of epistemic characterisation theorems, let me first examine the logic of decision-theoretic modelling, and then contrast it with the logic of game-theoretic modelling. Consider the decision problem shown in Figure 1.1 where agent i has to choose between actions i_1, i_2 and i_3, and where the possible states of nature are ω_1, ω_2 and ω_3. If i does not have any information about what state of nature will obtain, she may decide to act on a number of divergent and mutually inconsistent principles. She may decide to maximise the minimal payoff or utility, to minimise the maximum possible regret (risk), to use Hurwicz's pessimism-optimism index criterion, or she may use a principle of decision under uncertainty

---

[5] While 'Decision-Theoretic Foundations Programme', 'Epistemic Foundations Programme' or 'Interactive Epistemology' are good candidates, the terminology of 'Epistemic Programme' seems to be gaining popularity in the literature (Adam Brandenburger, personal communication).

[6] It may be worth pointing out that the ban on exogenous information still enjoys wide support. Adam Brandenburger and Amanda Friedenberg, 'Intrinsic Correlation in Games', *Journal of Economic Theory*, 141 (2008), 30 consider as endogenous the beliefs of the players in so far as they are derived from a hierarchy of common beliefs, emphasising that it is standard in the Epistemic Programme to consider these variables as part of the description of the game.

	ω_1	ω_2	ω_3
i_1	1	0	1
i_2	5	0	3
i_3	3	1	2

Fig. 1.1

based on Laplace's criterion of insufficient reason. But it is clear that she will never choose her first strategy—unless she wishes to lose—as it is *strictly* (or *strongly*) *dominated* by i_3.

Equally clear is that if she possessed information to the effect that the first or third state of nature will obtain, she would choose her second strategy and if she believed that the second state is the actual world, she would choose i_3. In easily understood formalism, this reasoning is of the form

$$(i\text{'s utility} \wedge i\text{'s rationality} \wedge \Box_i(\text{possible states of nature})) \to i\text{'s actions},$$

where the $\Box_i$ is used for i's beliefs. The sentence attests that i's actions ensue from her payoffs, rationality and beliefs about possible states of nature.[7]

	2_1	2_2	2_3
1_1	(1,5)	(0,1)	(1,3)
1_2	(5,0)	(0,0)	(3,1)
1_3	(3,3)	(1,1)	(2,2)

Fig. 1.2

Turning to game theory proper, consider the game shown in Figure 1.2. It is a normal form game between two players 1 and 2. Player 1 has the payoffs agent i had in the above decision problem, and as a result she will not play her first strategy

[7] I adopt the game-theoretic convention that actions and objects of beliefs are thought of extensionally. It is crucial to note the plural in the consequent, because rationality principles do not always fix unique actions.

here either. Furthermore, it is plain that her choice between her second and her third strategy ought to depend on what she believes about the prospective choices of her opponent. If he chooses 2_2, then 1_3 is the best choice for her; if he chooses something else, then she should play 1_2—unless, again, she wishes to lose.

The decision-theoretic analysis only proceeds if the agent possesses exogenous, statistical information about nature's prospective moves. How does the Epistemic Programme circumvent the use of such data? In the present case, player 1 can further develop her beliefs about her opponent in purely endogenous ways, that is, only referring to beliefs, utilities and rationalities. To see this, consider what player 1 can do with the extra information about player 2's utility function. Player 1 observes that playing 2_3 is always better for player 2 than playing 2_2, and this kind of reasoning helps her to decide upon her own action; playing 1_2 is the best choice if player 2 does not play 2_2. Nor is this all, because player 2 only avoids playing dominated strategies if he does not wish to lose. For that reason, player 1 also has to have information about her opponent's rationality; she should know that he maximises expected utility. In formalism, the assumptions allowing me to conclude that player 1 believes that player 2 does not play 2_2 and that therefore player 1 plays her second strategy are fully captured in the antecedent of

$$(\text{1's utility} \wedge \text{1's rationality} \wedge \Box_1(\text{2's utility}) \wedge \Box_1(\text{2's rationality})) \rightarrow \text{1's actions.}$$

It is natural to ask whether similar reasoning can be used to determine player 2's choice of strategy. The analogue of the above assumptions,

$$(\text{2's utility} \wedge \text{2's rationality} \wedge \Box_2(\text{1's utility}) \wedge \Box_2(\text{1's rationality})) \rightarrow \text{2's actions,}$$

is insufficient, though. Being rational, player 2 will not play 2_2; it is strictly dominated by 2_3. Strategy 2_3, moreover, is a better choice against 1_2, but 2_1 is better against 1_1 and 1_3. Since player 2 believes 1 to be rational, he will exclude her from playing 1_1. That does not help him too much, however, as long as he does not have more information about whether 1 will play 1_2 or 1_3. Unfortunately for player 2, there is no way to obtain more information on the basis of the epistemic setting described in the antecedent of the above implication. While player 1 considers 1_1 to be a bad strategy no matter what, her opinion about 1_2 and 1_3 depends on what she believes that player 2 will do. But as long as player 2 has no information about player 1's beliefs about what 2 will do, player 2 has no basis for beliefs about which of the two strategies 1_2 and 1_3 player 1 plays.

With more elaborate, yet still exclusively endogenous information, progress can be made, though. To see this, suppose that

$$\begin{aligned}\text{2's utility} \wedge \text{2's rationality} \wedge \Box_2(\text{1's utility}) \wedge \Box_2(\text{1's rationality}) \wedge \\ \Box_2\Box_1(\text{2's utility}) \wedge \Box_2\Box_1(\text{2's rationality}).\end{aligned}$$

Player 2 believes that 1 is rational and that 1 has a utility function as shown in Figure 1.2. From this, player 2 concludes that 1 will not play 1_1. Player 2 believes, in addition, that 1 believes that 2 is rational and that 2 has payoffs as stipulated in the game matrix. A rational player possessing such preferences never plays 2_2, and therefore player 2 believes that 1 believes that 2 will not play 2_2. Player 2 then observes that—provided 1 is rational and has the utility function the matrix details—the best response of 1 to the belief that 2_2 will not be played is playing 1_2. Consequently, player 2 believes that 1 will play 1_2. He has made his beliefs about his opponent more precise, and this allows him, in particular, to decide between 2_1 and 2_3. Believing that his opponent will play her second action, he chooses 2_3. That is, a precise description of player 2's choice of strategy can be derived from the more elaborate epistemic assumption.

The general form of epistemic characterisation results emerges. What I have established is something of the form

$$\begin{aligned}(\mathbf{u}_1 = \ldots) \wedge \mathbf{rat}_1 \wedge (\mathbf{u}_2 = \ldots) \wedge \mathbf{rat}_2 \wedge \\ \Box_1(\mathbf{u}_2 = \ldots) \wedge \Box_2(\mathbf{u}_1 = \ldots) \wedge \Box_1\Box_2(\mathbf{u}_1 = \ldots) \wedge \Box_2\Box_1(\mathbf{u}_2 = \ldots) \wedge \\ \Box_1(\mathbf{rat}_2) \wedge \Box_2(\mathbf{rat}_1) \wedge \Box_1\Box_2(\mathbf{rat}_1) \wedge \Box_2\Box_1(\mathbf{rat}_2) \rightarrow \mathbf{1}_2 \wedge \mathbf{2}_3,\end{aligned}$$

where $\mathbf{u}_i = \ldots$ abbreviates a complete description of i's utility function, $\mathbf{rat}_i$ means that i is an expected utility maximiser (is rational), $\mathbf{i}_k$ means the player i plays her kth strategy. Anticipating the discussion in Chapter 2, this is a particular instance of the epistemic characterisation theorem to the effect that common true belief about rationality and utility entails that players choose strategies that survive the *iterated elimination* of strictly dominated strategies. Epistemic characterisation results, in short, are sentences of the form

$$\varphi(\mathbf{rat}_1, \mathbf{rat}_2, \mathbf{u}_1, \mathbf{u}_2) \rightarrow \text{actions},$$

where φ is a formula in which epistemic operators $\Box_1$ and $\Box_2$ may be used, nested arbitrarily deeply. The statement about actions in the consequent is a statement of the form $\mathbf{1}_k \wedge \mathbf{2}_l$ describing a situation in which 1 plays her kth strategy and 2 his lth, or a (finite) disjunction of such statements $\bigvee_{k,l}(\mathbf{1}_k \wedge \mathbf{2}_l)$, if the antecedent epistemic conditions are insufficient to attribute performing one single action to each player.[8]

1.1.2 Normal Form Games

Different assumptions about beliefs, desires and rationality principles epistemically characterise different game-theoretic solution concepts.[9] Similar assumptions appear in almost every characterisation result, and the aim of this section is to expli-

[8] A consequence of non-uniqueness is that a full explanation of the specific action performed by an agent cannot always be given in entirely game-theoretic terms. Decision and game theory explain that the action actually chosen lies in some set of possible actions. But neither theory can always account for why one action is chosen rather than another.

[9] The treatment of these issues has benefited from detailed written comments by Wlodek Rabinowicz, for which I am very grateful.

cate them. While it is entirely false that no models could be developed for situations in which these assumptions are not satisfied, without them epistemic characterisation results would often make less sense to the Epistemic Programme in game theory—but, as will become clear, there are important exceptions. Evolutionary, behavioural, stochastic and cooperative game theory all provide ample space for relaxing the assumptions—and some non-cooperative game theorists may also wish to resist adopting the viewpoint promoted by the Epistemic Programme—but this does not contradict my argument in this section, because the aims of those researchers are crucially different from ours.[10]

To start with, players of a normal form game can choose from a set of possible actions, in most models finite, and never containing fewer than two elements. Modelling a situation as a decision or game theorist means ascribing to the agents weak total preference orderings over all possible outcomes of the game. Ordinal orderings are often sufficient in the Epistemic Programme, but the full force of the von Neumann–Morgenstern axioms is needed whenever preference orderings are to be uniquely represented by means of utility functions modulo linear transformations; this is the content of Theorem A.1 (see the Appendix A).[11] Games are almost never specified using preference orderings. Utility functions are the standard.

Even though preferences are essential, they are insufficient to give a full motivation for actions; they do not on their own constitute reasons to act, but stand in need of at least a principle telling the agent what to do with her preferences. Decisions under certainty or uncertainty allow for a fair number of different principles. Decisions under risk involve the maximisation of expected utility, and this is the principle employed in game theory, too.[12] In the Epistemic Programme, no agent ever acts without a principle of rationality.

It follows that acting is rationally choosing an action from a non-empty, non-singleton choice set. If agents did not choose, they would dawdle and leave the game unfinished, and if agents chose more than one action, they would be spoilsports and ruin the game. Only if all players choose precisely one action can the game reach an outcome. In normal form games, a related condition is often phrased by stipulating that players play simultaneously, which means that players have no chance to observe what their opponents do. They act in ignorance of what their opponents choose—notwithstanding the fact that they may have very accurate beliefs that would predict their opponents' actions. This requirement could be envisaged by players who make their choice in private first, handing their choice over to

---

[10] The stance here is logical, not epistemological. No critical evaluation of the plausibility of the assumptions is carried out; only an investigation into the logic underlying epistemic characterisation results.

[11] A weak total ordering is a reflexive linear ordering. To be precise, the ordering ranges over the lotteries composed of the possible outcomes of the decision problem or game, satisfying, in addition, conditions of monotonicity, substitutability, continuity and reduction, i.e., the von Neumann–Morgenstern axioms. See, e.g., Luce and Raiffa, op. cit. 23–31. For an alternative rendering that has become the standard in the Epistemic Programme, see the Appendix A.

[12] If a player's preference ordering is an ordinal one, or if her beliefs are not of the Kolmogorov form, not all of the mathematical details of expected utility maximisation are needed in full to make a player play, and a simpler definition of rationality can be used.

an objective umpire who reveals the choices—and thereby announces the achieved outcome—as soon as she has received them all.

Accompanied by highly specific utility functions, rationality principles are still often powerless if a player does not have beliefs. As I have suggested, fully pinning down the strategic choice of a game-theoretic agent involves quite complex reasoning with nested epistemic operators involving the beliefs of one player about those of another. In a typical situation, a player possesses beliefs about her and her opponents' possible actions, about her and her opponents' preference orderings, about her and her opponents' rationality principles, about her and her opponents' actions to be performed, and—with recursion intended to generate the necessary hierarchy of beliefs—about her and her opponents' beliefs.

Before turning to these five kinds of beliefs in more detail, it is important to realise that what applies to preferences also applies to beliefs: they take, in game theory, a specific kind of object. While in most concrete examples the beliefs as well as the preferences can be stated in everyday language, beliefs are regularly—but not always—taken to be probability measures over outcomes. They have to satisfy the Kolmogorov axioms to ensure, most importantly, that summing up probabilities is allowed whenever the probability of the disjunction of two independent events will be calculated. More intricate models are needed when we are interested in the ways players would change their beliefs if they learned that their current beliefs were wrong. Plain standard probability theory does not tell us much about how to apply Bayes' Rule to null events. Belief revision theory, by contrast, sorts out more (theoretically) rational from less rational ways to update and correct one's epistemic state. In the characterisation of several game-theoretic solution concepts, the Epistemic Programme has employed this theory to describe how the players' dispositions to revise beliefs influence the outcome of the game.

Of the five ingredients, the first is clear. Without beliefs about what to choose between, players cannot make a choice, and without beliefs about what outcomes may result, the players' choices of action cannot be guided by their opinions about the desirability of possible outcomes. Information about possible actions suffices to that end, because the set of possible outcomes is entirely determined by the possible actions of the players.

Second, players ought to have correct beliefs about their own preferences—or at least, beliefs that are sufficiently correct to guide decision making. If players believe they do not have preference orderings, preferences cannot be considered as reasons for their actions; and if they believe they have certain preferences, but their true preferences are very different, it is hard to tell whether they inform their agency. In fact, it seems as though completely incorrect beliefs about preferences are incoherent—at least for the purposes of decision and game theory. If players believe they have preferences that are different from those stipulated by the model, and act on those beliefs, then there is much to recommend that the theorist substitute the preference ordering in the model with the believed ordering. In the absence of any beliefs about preferences, only unconscious motivation would make sense, and without beliefs about preferences that are not at least approximately correct, there is no theory of games at all.[13]

This point is important, in particular if various degrees of autonomous choice are to be distinguished. Unconscious motivation has to be rejected, because players who are unconscious of their own preferences can hardly be thought to play a game. Of course, there is nothing incoherent about describing a person as gradually getting to know her real preferences and making conscious what she was previously unconscious of, and this may not be completely impossible in decision theory. In fact, someone's actions may be neatly modelled as maximin actions in some model ascribing preferences to the agent she first refused to acknowledge as her own. As the theorist continues modelling, however, we find that the model describes her behaviour correctly more often than not, and this prompts the agent to accept the model's specification of her preferences as correct. The models, one could say, explained her actions correctly, but it took a while before the agent herself became aware of that.

Yet this does not make sense in non-cooperative game theory as long as the perspective of the Epistemic Programme is adopted. A similar scenario where someone's actions are modelled as iteratively strictly undominated, for instance, may indeed turn out to describe her behaviour adequately, and the agent may indeed acknowledge that this reveals her true preferences. It is doubtful, however, whether a real explanation of her actions would be given, because the epistemic conditions of iterated strict dominance were not satisfied; these conditions involve common true belief about rationality and utility, and this was lacking as the modelled preferences were different from the actual preferences—the beliefs were common but not true. The distinction between players' *real* and *believed* preferences, in other words, is rejected, because at most only one preference ordering can play a motivating role, and the motivating preference ordering would need to appear in the game-theoretic model. Moreover, while not denying the conceptual possibility of players entertaining the belief that they do not possess any preference orderings, I reject the relevance of such possibilities to decision and game theory. If such beliefs are false, the preference ordering probably provides unconscious reasons for action; and if they true, there is just nothing decision and game theory can do.

Nor does it make sense, for game-theoretic purposes, to talk about players who are wholly uncertain about their preferences. If players are so uncertain about their preferences as to make it impossible to decide upon an action, then they just lack preference orderings. If, on the other hand, they can still decide on actions, they are, for game-theoretic purposes, just players with a particular preference ordering—or a range of possible orderings—that inspires the performance of one particular action. To summarise, as Adam Brandenburger and Robert Aumann write in their seminal paper on the epistemic characterisation of the Nash equilibrium, 'knowl-

[13] Players may be ignorant of their future preferences in extensive games under the many-moment interpretation.

edge of one's own payoff function may be considered tautologous'.[14] Without this assumption, the Epistemic Programme would not fully flourish.

Players not only need beliefs about their own preferences, they also need beliefs about the rationality principle that they use. If I explain the behaviour of a certain agent in terms of maximin, but she sees herself as solving a maximisation problem corresponding to the utility function and the probabilistic beliefs she believes herself to possess, then I misconstrue the reasons underlying her choice of action. Whenever I say that preferences, beliefs and rationality principles explain actions, I actually mean to refer to beliefs about preferences, beliefs about rationality and beliefs about a number of other aspects of the game-playing situation (available actions, outcomes, and so on).[15]

Of course, players have to have numerous beliefs about their own role in the game. The Epistemic Programme has helped to uncover the fact that players must have beliefs about each other. Games in which players do not have a clue about the preferences of their opponents are reducible, for them, to games against nature; all entries in the game matrix—except their own—can be removed, as far as they are concerned. Acting on principles from decision theory, they would choose on the basis of exogenous information. In the game shown in Figure 1.2 player 1 may form the belief that 2 will not play his second strategy on the basis of the information 1 has about 2's preferences, because player 1 observed that 2_2 is strictly dominated by 2_3. Without beliefs about 2's preference ordering and rationality, player 1 would not have been in the position to make the argument.[16]

In summary, for the Epistemic Programme in game theory to make sense, a player of a normal form game must have beliefs about her and her opponents' possible actions, about her and her opponents' preference orderings, about her and her opponents' rationality principles, about her and her opponents' actions to be performed, and, with recursion to generate the necessary hierarchy of beliefs, about her and her opponents' beliefs. This five-tiered structure reoccurs in game-playing situations of extensive games—but with crucial complications.

---

[14] They state that 'Knowledge of one's own payoff function may be considered tautologous', 'Epistemic Conditions for Nash Equilibrium', *Econometrica*, 63 (1995), 1162. This is not to exclude Bayesian games and other models of incomplete or imperfect information. The aim of modelling such games is to capture situations where individuals are less than fully informed about the preference relations of their opponents, or even about certain exogenous features of the actual world. This is compatible with the claim I defend, because—to stay within the framework of the Epistemic Programme—reasoning about solution concepts of such games involves considering larger games in which these pieces of information have been 'endogenised'. For a textbook treatment, see Martin Osborne and Ariel Rubinstein, *A Course in Game Theory* (Cambridge: MIT Press, 1994), 24–27.

[15] I use the concept of *game-playing situation* rather loosely. Cf., e.g., Adam Brandenburger, 'The Power of Paradox: Some Recent Developments in Interactive Epistemology', *International Journal of Game Theory*, 35 (2007), 465–492.

[16] Again, these assumptions can be somewhat relaxed. Sensible conclusions can be derived about the likelihood of opponents performing certain strategies even when a player only has approximate, not entirely accurate beliefs about opponent utility functions. For further discussion see Section 2.4.

1.1.3 Extensive Games: The One-Shot Interpretation

Extensive games model temporally extended, sequential strategic interaction. Normal form games, by contrast, model one-shot events. That, at least, is the common view. For von Neumann and Morgenstern, however, normal form and extensive form games are different possible models of one and the same sort of thing. 'Imagine,' they write,

> that each player... instead of making each decision as the necessity for it arises, makes up his mind in advance for all possible contingencies.[17]

Then this is no restriction of his freedom of action

> because the strategy is supposed to specify every particular decision only as a function of just that amount of actual information which would be available for this purpose in an actual play.[18]

One thing can be modelled in two different ways. The *normalised* form is more appropriate for proving general results, the *extensive* form is better when one wishes to analyse particular cases, but for the founding fathers of game theory the two forms are 'strictly equivalent'.[19] To analyse the characterisation results that the Epistemic Programme has obtained about extensive form games, it pays to set out these two interpretations more clearly, and I do this by distinguishing between a one-shot interpretation and a many-moment interpretation of extensive form games. The latter interpretation is most sensitive to sequentiality, but the one-shot interpretation, too, is still different in subtle respects from pure normal form games.

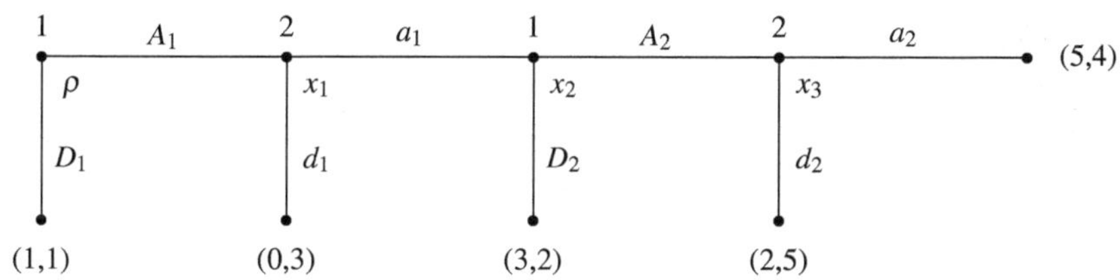

Fig. 1.3

According to the one-shot interpretation, players of an extensive game act simultaneously, and choose between a set of strategies fixing the actions at any of their

[17] Op. cit. 79

[18] Ibid.

[19] Ibid., 85. It is not terribly clear what 'strict equivalence' means here. For further discussion see Boudewijn de Bruin, 'Game Transformations and Game Equivalence', ILLC Technical Note X-1999-01 (University of Amsterdam, 1999), and Susan Elmes and Philip Reny, 'On the Strategic Equivalence of Extensive Form Games', *Journal of Economic Theory*, 62 (1994), 1–23.

decision nodes in the game. In the Centipede game shown in Figure 1.3, for instance, player 1 can choose between the strategies D_1D_2, D_1A_2, A_1D_2 and A_1A_2. In order to find the assumptions on one-shot extensive game-playing situations one could expect, first, that copying the conditions on the five elements of normal form game-playing situations suffices. Preference orderings surely satisfy the von Neumann–Morgenstern axioms, players act on principles of rationality, and they ought to have beliefs about possible strategies, performed strategies, preferences, rationality principles, and, with the usual recursion, about beliefs. Moreover, similar to playing normal form games, players decide on precisely one *full strategy* (a function, here, mapping any decision nodes of hers to an immediate successor) and choose simultaneously.

Under the one-shot interpretation, players of an extensive game can choose between a set of possible full strategies prescribing a unique action at any decision node. At every decision node more than one action is possible, and on that account the set of possible full strategies contains at least two elements. From these possible strategies, players pick precisely one, without the game structure allowing them to have information about the choices that the others make. In short, they choose simultaneously.

Preference orderings in normal form game-playing situations range over strategy profiles combining the strategic choices of all the players. Different strategy profiles give rise to different outcomes with different—or identical—utility for one or more players. Preference orderings in one-shot extensive game-playing situations, by contrast, are defined over the terminal nodes of the game tree, the genuine outcomes of the game, and the same terminal node can often be reached by more than one full strategy. In the Centipede , for instance, strategy profiles (A_1D_2, d_1d_2), (A_1A_2, d_1d_2), (A_1D_2, d_1a_2) and (A_1A_2, d_1a_2) determine the same outcome.

More subtle issues arise when we consider rationality in the one-shot interpretation, as we have to decide whether one-shot principles of rationality differ from normal form principles. Using normal form rationality as one-shot rationality, a full strategy in an extensive game is rational for a player given her beliefs and utility function whenever the full strategy is rational in the normal form version of the game. Different notions of rationality for normal form games (weak domination, strict domination, perfect rationality, and so on) will give rise to different notions of rationality for extensive games, without, however, essentially referring to their sequential character. An alternative way to define one-shot extensive game rationality does exploit the extra structural properties of extensive games by considering a full strategy not only in the whole game but also in all of its subgames. Imagine, for instance, that player 2 in the Centipede wishes to quit the game at the first decision node, because this is presumably the best thing to do given the player's beliefs and preferences. Looking at the normal form of the game, it does not matter whether she plays d_1d_2 or d_1a_2, because the outcome will be the same with either of these two full strategies. The latter prescribes a bad action in the subgame generated by x_3, though, for player 2 would lose one unit of utility in comparison to playing d_2. That being the case, a notion of rationality sensitive to subgames excludes player 2 from playing d_1a_2 without, it is important to note, transcending the one-shot framework—it only involves player 2's one-time event of choosing a full strategy.

If rationality principles come in two versions, it might be expected that beliefs come in a subgame-insensitive and a subgame-sensitive version also. The subgame-insensitive conception takes the full strategies of players' opponents as objects of their beliefs. In order to get some impression of the ways a subgame-sensitive conception can be defined, consider the ways that player 1's belief that 2 will quit the Centipede game as soon as possible can be phrased. Since the one-shot interpretation holds to the view that the only objects of choice are full strategies, player 1's beliefs should be about 2 playing d_1d_2 or d_1a_2. She could believe that player 2 plays the former, she could believe he plays the latter, and she could believe that either he plays the former or he plays the latter, but in all three cases she believes that at x_1 player 2 quits the game. For the rational evaluation of player 1's choice of a one-shot full strategy makes no difference which of the three beliefs she uses, because all that matters is what she does at the root of the game. In order to evaluate a full strategy in, say, the subgame generated by x_2, it matters what exactly she expects of her opponent in that subgame.

There are two ways out. First, players may be supposed to have beliefs for all subgames. Player 1 may believe that player 2 will play d_1a_2 in the entire game, but that in the subgame generated by x_2 he will play d_2, for instance. Alternatively, we could presuppose that a player's beliefs about the *entire* game are serious. If a belief entails a certain action at a certain node which, according to that very belief, will not be reached, the belief about this action still expresses serious reasoning about the opponent's strategic situation. Player 1 should, following this latter logic, believe that 2 will play d_1d_2 in the entire game, and that belief would be reusable in the subgame generated by x_2 stating that 1 believes 2 to play d_2 at x_3.

I do not have a strong opinion about which of the two is the best, because, I believe, they are both antithetical to the one-shot interpretation of extensive game-playing. In a one-shot game-playing situation, players make up their minds about full plans of action, without envisaging any future moments of decision. Players who do not envisage future moments at which they can decide, do not envisage future moments at which they have beliefs either, and if I am right here, special subgame-sensitive beliefs, under the one-shot interpretation, do not make sense. As a consequence, rationality principles which pay attention to subgames do not make sense according to the one-shot interpretation for the same reason. Beliefs about subgames and rationality, in other words, are about what would happen if subgames were reached at some later point in time. But under the one-shot interpretation it does not make sense to consider such possibilities. Subgames are not reached at all.

Ultimately, the one-shot interpretation comes in only one version—a radical one. Playing an extensive game in the one-shot interpretation is playing its normal form version without rationality principles that go beyond the normal form, and Chapter 3 discusses important consequences for the epistemic characterisation results adopting the one-shot view.

1.1.4 Extensive Games: The Many-Moment Interpretation

The one-shot interpretation seems to stay closest to von Neumann and Morgenstern's dictum that normal form and extensive form games are equivalent. To motivate considering a many-moment interpretation of extensive game-playing, a possible counterargument against the equivalence would hold that a player who thinks of the game as consisting of various subsequent decision moments will just not develop a full plan of action for the game—and hand it to the umpire overseeing it—but will rather think only about the choice of action to be made at the decision node she thinks she is at. The many-moment interpretation indeed conceives of a game-playing situation as a sequence of decision moments. Game-play, that is, entails a run through the game tree terminating in some outcome, but it entails more, because every decision moment ought to contain the necessary elements to explain the action performed at any decision moment in the run—preferences, principles and beliefs. A game-playing situation is accordingly a run through the game tree plus some extra information.[20]

The essential difference between the one-shot and the many-moment interpretations is that the former has players choose a full strategy at one single point in time, while the latter sees players as making up their minds at various points in time. In the many-moment interpretation, players may exert influence on the outcome more than once, respond to their opponents' moves, and block entire subgames by their choices of action. Since a many-moment game-playing situation is a sequence of decision moments, the logical form of the various components of decision moments is different from the corresponding element of the one-shot interpretation. At every decision moment a player has to act rationally on the basis of some preferences and beliefs. A player's beliefs may, first, be thought of as relating all of her possible actions at some decision node to the terminal nodes she expects to ensue if she were to choose that action, and however minimal this conception of beliefs may be, it is enough to make rationality principles work: choose the action you expect to lead to the best possible terminal node. Yet as it stands this conception of beliefs is in almost all cases unserviceable to the Epistemic Programme's purposes, because only very occasionally will the utility structure of the game be sufficient for the players to base their expectations on. Outcomes that are worst for everyone may be excluded, and those that are best for everyone expected, but such cases are rare, and statistical data and the like are clearly ruled out by the ban on exogenous information.

As long as the players' beliefs concern terminal nodes, they may be said to possess beliefs about a restricted number of possible eventualities only, and I therefore propose to add more structure to beliefs to make them fit for the many-moment

---

[20] Another interpretation comes into sight when we consider players deciding on the performance of entire strategies, but not necessarily once and for all. Such an interpretation is mentioned by Wlodek Rabinowicz, 'Grappling with the Centipede: Defence of Backward Induction for BI-terminating Games', *Economics and Philosophy*, 14 (1999), 99. See also ibid., 'To Have One's Cake and Eat It, Too: Sequential Choice and Expected-Utility Violations', *Journal of Philosophy*, 92 (1995), 586–620.

interpretation. In this interpretation, players have beliefs about all possible future decision nodes, giving rise to an expected path through the game and its subgames.

Apart from beliefs about choices of action, players have beliefs about possible actions, preferences, principles, and, with recursion, beliefs, plus—new for the many-moment interpretation—beliefs about the actions played prior to the current decision moment. Beliefs about possible future actions and beliefs about previous actual actions are beliefs about the tree structure of the game and the position therein. Beliefs about preferences are beliefs about the ordering over terminal nodes.

Formal renderings of those beliefs, as well as the even more obvious beliefs about beliefs, may disagree on the details of the logical form, but these issues need not detain me here. What should be discussed, however, are the different ways in which the many-moment interpretation can deal with the history of game-play. The one-shot interpretation has players form beliefs about their opponents' choice of full strategies. The many-moment interpretation has players form beliefs about their opponents' choices of action at each and every possible decision node of theirs.[21] Now, imagine a player at some decision moment who wishes to form a belief about what is going to happen at some future decision node. If it is her own decision node, she could simply commit herself to performing some action there, but that would be strikingly incoherent with the many-moment interpretation, because she would in that case adopt a full strategy as in the one-shot case. To be a genuine many-moment agent, she should imagine herself in the particular future decision node, find out what preferences, principles and beliefs she will act upon then and there, and deduce from that a belief about her own choice of action. This is no different from the way players form beliefs about each others' prospective choice of action at future decision nodes. Beliefs about future preferences, principles and beliefs, that is, are used to deduce predictions of future actions.

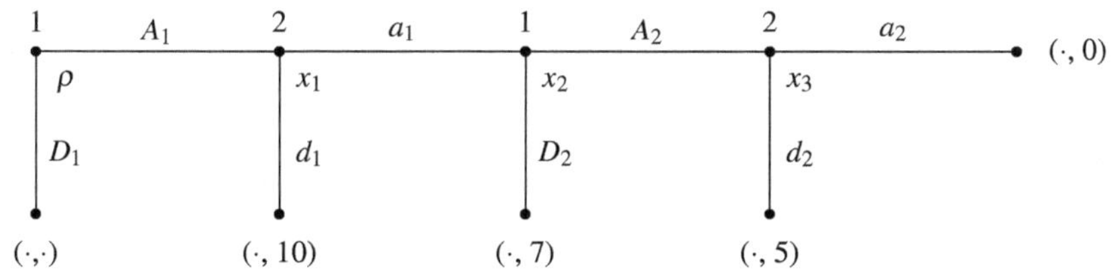

Fig. 1.4

[21] In the one-shot interpretation, beliefs may also involve subgames and in the many-moment interpretation, beliefs may also be about terminal nodes. As I have shown, however, these are either insufficient to run the Epistemic Programme, or incoherent given the further details of the interpretation.

How can players obtain any information about these elements? As with normal form games and the one-shot interpretation, general conditions on game-playing situations such as common true belief about rationality and utility may be used. A new source of information springs, however, from the decision nodes and their positions in the game tree. Consider, for instance, a variant of the Centipede, shown in Figure 1.4, of which only the relevant parts of the utility functions have been drawn. Player 1 at the root of the game wants to predict what is going to happen at x_3, and tries—in a first attempt—to find out what 2's preferences, principles and beliefs will be at x_3. She may reason that player 2 is rational and that going down at x_3 is the only rational move whatever player 2 believes, concluding that at x_3 player 2 will play d_2. She could also—a second attempt—argue that imagining player 2 at x_3 entails imagining player 2 having chosen a_1 at x_1. Playing a_1 at x_1 is, player 1 believes, not rational for player 2, whatever his beliefs are, and therefore imagining player 2 at x_3 entails imagining player 2 with a past of irrational play. Player 1 may go on to note that there are various ways to make sense of x_3 being reached. Player 2 may have made a mistake at x_1, he may be genuinely irrational, and so on, and depending how player 1 continues the image of player 2 at x_3 different expectations about his action at x_3 follow. If that is right, without extra assumptions it is not clear what beliefs player 1 forms.

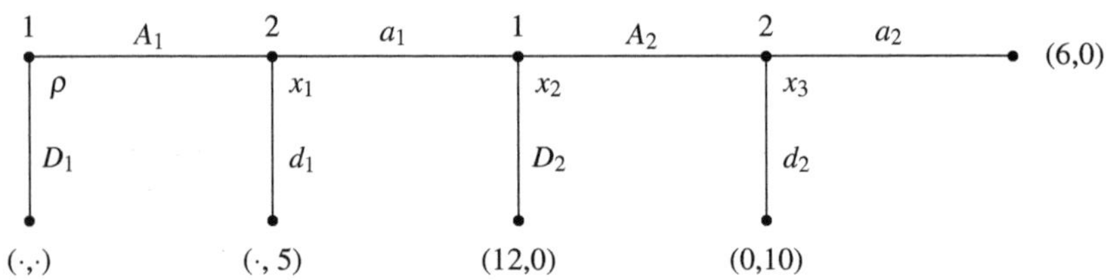

Fig. 1.5

The example suggests that players who take a *history-sensitive* look at decision nodes are in a worse position than those who adopt a *history-insensitive* view, because history often involves more than one, possibly conflicting explanation. To reinforce this point, consider yet another Centipede, shown in Figure 1.5. Player 1 at the root of the game again wants to know what is going to happen at x_3. If she takes the history of x_3 seriously, she has to consider 2's choice of a_1 at x_1. As a_1 is rational only if 2 expects 1 to play A_2 at x_2, picturing x_3 as being reached means picturing not only 2's action and principle at x_1 (as in the previous example) but also 2's beliefs at x_1. In order to form a belief about what is going to happen at x_3, it matters what 1 believes about 2's beliefs at x_1; it matters, for instance, to the question whether 1 ought to consider x_3 the result of mistakes, irrational play, or otherwise.

In summary, the most far-reaching consequences of the many-moment interpretation of extensive game-play pertain to beliefs, partly because of their content (terminal nodes, or actions at all decision nodes of the subgame) and partly because of their formation (history-sensitivity, or not). I will briefly examine assumptions that the many-moment interpretation places on further elements of game-playing situations.

A weak total ordering of the terminal nodes representable by a von Neumann–Morgenstern utility function captures the preferences of the players, and the distinction between strictly and weakly dominated actions can be used here, too, to formalise different concepts of rationality. Equally clearly, under the many-moment interpretation players of an extensive game choose from a set of possible actions at any decision node on the path followed through the game tree. Yet while the agent who moves at some decision moment of a game-playing situation chooses exactly one action among the possible actions, the many-moment interpretation has players play in turn, not simultaneously; exactly one player plays at a decision moment.

At a decision moment, the player who moves has to know where she is in the game tree. This general condition entails that she knows what has happened up to that moment, which actions have been chosen and which ones have been disregarded. She needs to know the entire past and every possible future of the entire past. This condition further entails that the player knows which actions she can choose. She needs to know every possible future of every possible action and by the same token she also has to know what her opponents can do at later decision moments. If these conditions do not hold, no genuinely game-theoretic account can be given of her actions, because particular features of the game theorist's model (for instance, particular preference orderings over terminal nodes of some subgame) would not play a role in the agent's own motivation for her actions.[22] In addition, a player should know her own preferences and those of her opponents, and she should know on which principle she acts at the decision node she is at.[23]

It is clear that without any beliefs about the opponents' rationality principles at possible subsequent decision moments the player cannot form any argument about the expected outcome—unless she reasons on the basis of statistical data and other forms of information that were excluded by the ban on exogenous information. However, how one sees such belief formation depends on how one sees decision nodes—with or without appreciation of their histories. The obvious requirement for the history-insensitive view is that players ought to believe that their opponents are rational at every possible decision node. There does not seem to be an obvious requirement for the history-sensitive conception.

Players should know what they play at a decision node and they should know what they have played (a consequence of earlier conditions, called *perfect recall*)

---

[22] It may be that these conditions will often fail to hold. What concerns me here, however, is not their conceptual plausibility or empirical adequacy. Rather, I indicate here what makes an explanation a game-theoretic one.

[23] This does not mean that she has to know the principles to be applied at any possible (or even subjectively probable) future decision node. For a discussion of weakening knowledge about opponents' utility functions, see Sections 2.3 and 2.4.

but they should not be required to know what they will play at possible future decision moments, because that would approximate the one-shot interpretation. Past actions of the opponent, however, ought to be known (a consequence of earlier requirements, called *perfect information*).[24] Yet players do of course generally have beliefs about future actions and about beliefs about future actions, for without beliefs about future beliefs a player would not be able to form any beliefs about what actions she would choose at that future decision node. Since she cannot, under the many-moment interpretation, truly commit herself to some future action already, she must form a belief about her future actions on the basis of beliefs about her future preferences, principles and beliefs. Similarly, beliefs about opponents' future beliefs are needed. Beliefs about past decision nodes, however, seem only required under the history-sensitive view. [25]

1.1.4.1 Identity Over Time

This admittedly rather rote discussion shows how conditions required for normal form game-playing situations as well as for the one-shot interpretation of extensive games can be adapted to the specific characteristics of many-moment game-playing situations. It is not all old wine in new bottles, though. I will now turn to conditions without earlier analogues, and ask how the five components of decision moments (possible actions, preferences, principles, beliefs and choices of action) evolve over time. Is it necessary to assume conditions relating those ingredients at different decision moments? Such a question does not of course make sense for possible actions, preferences and choices of actions, because if these ingredients changed the game-theoretic model proposed by the theorist would not be the game really played. But for principles and beliefs the question is more than pertinent.

In epistemic characterisation theorems game-theoretic explanations follow decision-theoretic explanations in that actions are taken to result from rational choice, and as there is more than one such principle, players may change their rationality principles over time. A player may start playing only strictly undominated actions and gradually come to play weakly undominated ones too, for instance. Such changes may be relatively uncommon, but similar changes in beliefs are, of course,

---

[24] This does not mean that games with imperfect information or imperfect recall are neglected here. The way I treat such games in Chapter 5, however, follows the one-shot interpretation and the reason is that only if players can foresee future informational asymmetries can a genuinely game-theoretic explanation be given. Overstating it slightly, games with imperfect information and games with imperfect recall do not make sense under the many-moment interpretation.

[25] Players have beliefs about the future, which may be justified and even true, but in the explanation of the players' actions the theorist cannot go beyond belief and make essential use of the fact that the beliefs, in fact, constitute knowledge. This also applies to normal form and one-shot situations. This does not mean that knowledge (as opposed to mere belief) never adds to the explanation of an action. Timothy Williamson, *Knowledge and its Limits* (Oxford: Oxford University Press, 2000), 60–64 gives an argument in favour of the 'causal efficacy' knowledge may have for human agency. That aspect of a knowledge constituting belief that makes it causally efficacious (something like its justification) has not been dealt with in the Epistemic Programme in game theory.

entirely natural. In fact, (theoretic) rationality entails that several beliefs change, most notably beliefs about the position in the game. It is useful to distinguish between the beliefs a player has about herself and beliefs she has about her opponents, because players have introspective access to aspects of the (possible) development of their agential identity that they do not have about their opponents. Beliefs about the players' own pasts will not generally change, as they know on which preferences, principles and beliefs they acted, but beliefs about their possible futures may change quite radically. Suppose that in some extensive game at x_k you believe that at x_m you will play on a principle of rationality. At some intermediate decision moment x_l of yours, however, you may feel differently about that situation. Introspection may tell you that your attitude to the game has changed—which does not necessarily mean that your preference ordering over terminal nodes has changed—and this may make it less likely that you will act on a principle of rationality at x_l. This does not strike me as a very common phenomenon of special use to game-theoretic modelling, but it does not seem incompatible with the view of action explanation underlying the Epistemic Programme. So, beliefs about your possible future principles may change, and beliefs about your beliefs about possible principles may change also.

Similarly, a player's beliefs about her opponent's past principles and beliefs may change. At x_l you may believe that your opponent's action at some preceding x_k was the result of a mistake. At some later x_m you may have seen more of his decisions, and this new information may force you to reconsider your earlier beliefs, but these beliefs are irrelevant in terms of future decision making. However, beliefs about the future principles and beliefs of a player's opponents ought not to change. If we adopt a history-insensitive view of decision nodes, picturing future decision nodes is independent of any information about previous play; and if we adopt a history-sensitive view of decision nodes, the picture of future decision nodes already takes care of all possible information about past play—as counterintuitive as it may sound.

1.2 A Logic for Game Theory

As I have shown, the research presented in this book studies game theory from two perspectives. It contributes internally to the Epistemic Programme in game theory itself by developing an epistemic logic for game theory, and it criticises externally the applicability of game theory as a descriptive and normative endeavour from the point of view of epistemology.[26]

The conceptual study of normal form and extensive game-playing situations as the Epistemic Programme views it cannot be used to derive stable results as long as no appropriate formalism is available to capture what would otherwise remain tacit and imprecise. I will now present the bare bones of the formalism. In the next two chapters, I will use the formalism to represent a number of existing results from the

---

[26] I have articulated my views on the interrelations between epistemic logic and epistemology in Boudewijn de Bruin, 'Epistemic Logic and Epistemology', in V. Hendricks and D. Pritchard (eds.), *New Waves in Epistemology* (Basingstoke: Palgrave Macmillan, 2008), 106–136.

Epistemic Programme, both to enable better and more precise comparisons, and to obtain new epistemic characterisation results. The last two chapters will then refer back to the formal results and use them in their critical evaluation of descriptive and normative game theory as well as in a comparison of the Nash Equilibrium Refinement Programme and the Epistemic Programme.[27]

1.2.1 A Logic for Normal form Games

Given an N-person normal form game with multi-matrix $(p_{i,k_1,\dots,k_N})_{i,k_1,\dots,k_N}$ representing the utility structure, I will define a formal language to describe all aspects of game-playing that are relevant to the study of the epistemic and rationality assumptions underlying game-theoretic solution concepts.[28]

The logical symbols used are $\neg$ (*not*, negation), $\wedge$ (*and*, conjunction), $\vee$ (*or*, disjunction), $\rightarrow$ ('if..., then...', implication), and $\leftrightarrow$ ('...if and only if...', equivalence). No quantifiers $\forall$ ('for all') or $\exists$ ('there exists') are needed. The conjunction (disjunction) of all sentences from a finite set Σ is abbreviated by $\bigwedge\Sigma$ ($\bigvee\Sigma$), assuming commutativity. If the φ_i enumerate Σ we may also write $\bigwedge_i \varphi_i$ ($\bigvee_i \varphi_i$).

In order to obtain a genuine modal system, the set of logical symbols is enlarged with three operators. Depending on the precise proof system, the $\Box$-operator (historically with interpretation 'it is necessary that...') has a *doxastic* reading ('it is believed that...') or an *epistemic* reading ('it is known that...'). Since we are dealing with more than one player it makes sense to index the operators $\Box_i$ for each player i. The *dual* of the $\Box_i$ is written $\Diamond_i$; that is, $\Diamond_i$ abbreviates $\neg\Box_i\neg$. The $\mathbf{E}_I$-operator stands for 'every player $i \in I$ believes (knows) that...', and the $\mathbf{C}_I$-operator is used to speak about *common* belief (knowledge)—all players believe (know)..., and all players believe (know) that all players believe (know)..., and all players believe (know) that all players believe (know) that all players believe

[27] The Epistemic Programme has benefited from an inspiring number of studies in logic and games, and the present framework, while original in its formalisation of rationality in characterisation of iterated dominance solution concepts, is indebted to the work of various authors including Johan van Benthem, 'Games in Dynamic-Epistemic Logic', *Bulletin of Economic Research*, 53 (2001), 219–248, ibid., 'Extensive Games as Process Models', *Journal of Logic, Language and Information*, 11 (2002), 289–313, Oliver Board, 'Dynamic Interactive Epistemology', *Games and Economic Behavior*, 49 (2004), 49–80, Thorsten Clausing, 'Doxastic Conditions for Backward Induction', *Theory and Decision*, 54 (2003), 315–336, ibid., 'Belief Revision in Games of Perfect Information', *Economics and Philosophy*, 20 (2004), 89–115, Aviad Heifetz and Philippe Mongin, 'Probability Logic for Type Spaces', *Games and Economic Behavior*, 25 (2001), 31–53, Graham Priest, 'The Logic of Backwards Inductions', *Economics and Philosophy*, 16 (2000), 267–285, Robert Stalnaker, 'On the Evaluation of Solution Concepts', *Theory and Decision*, 37 (1994), 49–73, ibid., 'Knowledge, Belief and Counterfactual Reasoning in Games', *Economics and Philosophy*, 12 (1996), 133–163 (repr. with proofs in C. Bicchieri, R. Jeffrey and B. Skyrms (eds.), *The Logic of Strategy* (New York: Oxford University Press, 1999), 3–38), ibid., 'Belief Revision in Games: Forward and Backward Induction', *Mathematical Social Sciences*, 36 (1998), 31–56, and ibid., 'Extensive and Strategic Forms: Games and Models for Games', *Research in Economics*, 53 (1999), 293–319.

[28] For further discussion of notations, definitions and some theorems, see the Appendix A.

(know)..., and so on *ad inf.* An abbreviation for $\mathbf{E}_I \dots \mathbf{E}_I\varphi$ with n occurrences of $\mathbf{E}_I$ is $\mathbf{E}_I^n\varphi$. Furthermore, $\mathbf{E}_I\varphi \wedge \mathbf{E}_I^2\varphi \wedge \dots \wedge \mathbf{E}_I^n\varphi$ is written $\mathbf{E}_I^{\leq n}\varphi$. This is referred to as common belief (knowledge) *up to level n*. Probabilistic expressions $\mathbf{P}_i(\cdot) = \cdot$ represent i's probabilistic beliefs and arbitrary finite sums of such expressions $\mathbf{P}_i(\varphi_1)\cdot\mathbf{q}_1 + \dots + \mathbf{P}_i(\varphi_n)\cdot\mathbf{q}_n \geq \mathbf{q}$ are allowed as long as they are not mixed over players (as $\mathbf{P}_i(\varphi_1)\cdot\mathbf{q}_1 + \mathbf{P}_j(\varphi_2)\cdot\mathbf{q}_2 \geq \mathbf{q}$ would be for $i \neq j$), and obvious abbreviations use Σ.

The non-logical symbols include proposition letters to speak about games. Proposition letters $\mathbf{i}_m$ stand for the statement 'i plays her mth strategy i_m'. The formal analogue of the statement that $u_i(1_{k_1},\dots,N_{k_N}) = r$ for some real number r is $\mathbf{u}_i(k_1,\dots,k_N) = \mathbf{r}$. In order to be able to make all relevant statements involving utility, we need countably many symbols to refer to the real numbers (never all, sometimes finitely many). Rationality conceptions, finally, correspond to proposition letters of the form $\mathbf{meu}_i$, $\mathbf{rat}_i$, $\mathbf{prat}_i$ and $\mathbf{mrat}_i$, which I will explain later.

For modal logics, a Hilbert-style proof system is common and convenient. A proof in such a system of a sentence φ from a set of sentences Σ is roughly a finite sequence of sentences that are either taken from Σ, or axioms (typical for the particular Hilbert system), or statements derived (by rules typical for the particular system) from sentences occurring earlier in the sequence, such that the last sentence is φ. If such a sequence exist—if φ is derivable from Σ—one writes $\Sigma \vdash \varphi$. Generally, the axioms contain all modal or non-modal instances of tautologies from (classical) propositional logic. No extra creativity is needed to derive statements of the form $\varphi \vee \neg\varphi$ or $\varphi \rightarrow (\psi \rightarrow \varphi)$.

For the part of the language concerned with modality the following axioms are needed.[29] They are routinely stated for completeness and future reference.

Prop All classical propositional tautologies.
Dual $\Diamond_i\varphi \leftrightarrow \neg\Box_i\neg\varphi$.
K $\Box_i(\varphi \rightarrow \psi) \rightarrow (\Box_i\varphi \rightarrow \Box_i\psi)$.
T $\Box_i\varphi \rightarrow \varphi$.
D $\Box_i\varphi \rightarrow \Diamond_i\varphi$.
4 $\Box_i\varphi \rightarrow \Box_i\Box_i\varphi$.
5 $\Diamond_i\varphi \rightarrow \Box_i\Diamond_i\varphi$.
E $\mathbf{E}_I\varphi \leftrightarrow \bigwedge_i \Box_i\varphi$.
C $\mathbf{C}_I\varphi \leftrightarrow \mathbf{E}_I(\varphi \wedge \mathbf{C}_I\varphi)$.

In proof systems including the E-axiom every axiom for the $\Box_i$ is provable for the $\mathbf{E}_I$.[30] For instance, if the T-axiom and the E-axiom are available, all instances of $\mathbf{E}_I\varphi \rightarrow \varphi$ can also be appealed to. In proof systems including the C-axiom as well as the rule of induction from below, every axiom for the $\Box_i$ is provable for the $\mathbf{C}_I$.

29 Not all axioms are always needed, and this feature, highlighted in the discussion of the epistemic characterisation results in the next two chapters, is conceptually as well as technically quite interesting.

30 The Dual-axiom and the E-axiom are, in some way, not genuine axioms, but definitions of operators. This does not apply to the C-axiom, because the **C**-operator cannot be defined in the finitary language applied here.

Taking $\Box_i$ as the epistemic modality, the K-axiom expresses that what is known to be a logical consequence of something known is known; the epistemic subject i is *logically omniscient*. The Dual-axiom fixes the meaning of the $\Diamond_i$-operator. The T-axiom captures *veridicality*; what is believed is true. The D-axiom states the *consistency* requirement that what you believe is not believed not to hold. The 4-axiom formalises *positive introspection* of doxastic or epistemic states; you know that you know what you know. The 5-axiom, in turn, formalises that you know that you do not know something if you know not to know it—*negative introspection*. The E-axiom determines that the $\mathbf{E}_I$-operator expresses the beliefs everyone has. The C-axiom, finally, captures rather cryptically what it means that something is commonly known: everyone knows it, everyone knows that everyone knows it, and so on *ad inf.*

To discuss linear (in)equalities (to do the calculation necessary to solve maximisation problems) the following axioms are needed.[31]

0-term $\Sigma_k \mathbf{P}_i(\varphi_k)\cdot\mathbf{q}_k \geq \mathbf{r} \leftrightarrow \Sigma_k \mathbf{P}_i(\varphi_k)\cdot\mathbf{q}_k + \mathbf{P}_i(\varphi_{k+1})\cdot 0 \geq \mathbf{r}$.
Per $\Sigma_k \mathbf{P}_i(\varphi_k)\cdot\mathbf{q}_k \geq \mathbf{r} \leftrightarrow \Sigma_k \mathbf{P}_i(\varphi_{l(k)})\cdot\mathbf{q}_{l(k)} \geq \mathbf{r}$ for l any permutation.
AddCoef $(\Sigma_k \mathbf{P}_i(\varphi_k)\cdot\mathbf{q}_k \geq \mathbf{r} \wedge \Sigma_k \mathbf{P}_i(\varphi_k)\cdot\mathbf{q}'_k \geq \mathbf{r}') \rightarrow \Sigma_k \mathbf{P}_i(\varphi_k)\cdot(\mathbf{q}_k+\mathbf{q}'_k) \geq (\mathbf{r}+\mathbf{r}')$.
MultCoef $\mathbf{c} \geq 0 \rightarrow (\Sigma_k \mathbf{P}_i(\varphi_k)\cdot\mathbf{q}_k \geq \mathbf{r} \leftrightarrow \Sigma_k \mathbf{c}\cdot\mathbf{P}_i(\varphi_k)\cdot\mathbf{q}_k \geq \mathbf{c}\cdot\mathbf{r})$.
Dich $t \geq \mathbf{r} \vee t \leq \mathbf{r}$ for t any term.
Mon $\mathbf{q} > \mathbf{r} \rightarrow (t \geq \mathbf{q} \rightarrow t > \mathbf{r})$ for t any term.

To allow for probabilistic reasoning the Kolmogorov axioms are essential.

NonNeg $\mathbf{P}_i(\varphi) \geq 0$.
True $\mathbf{P}_i(\top) = 1$.
False $\mathbf{P}_i(\bot) = 0$.
Add $\mathbf{P}_i(\varphi) = \mathbf{P}_i(\varphi \wedge \psi) + \mathbf{P}_i(\varphi \wedge \neg\psi)$.
Dist $\mathbf{P}_i(\varphi) = \mathbf{P}_i(\psi)$ whenever $\varphi \leftrightarrow \psi$ is a propositional tautology.

In order to ensure that probabilistic and non-probabilistic beliefs are related in the right way, two additional axioms are useful.

Cons $\Box_i\varphi \leftrightarrow \mathbf{P}_i(\varphi) = 1$.
KnProb $\varphi \rightarrow \Box_i(\varphi)$ for φ an i-probability sentence.

The first of these interrelation axioms is a consistency requirement on the relation between non-probabilistic and probabilistic beliefs. In a sense it guarantees the maximum possible, showing that as the $\Box_i$ become redundant, the investigations can in principle be carried out in a stricter language. A weaker interrelation axiom can be defended by demanding an implication in the direction from left to right only. But since it is not necessary here either to allow for cases in which a player holds some proposition φ possible without assigning positive probability to it (or the converse of this case) or to distinguish between beliefs with probability one and non-probabilistic beliefs, I will ignore this subtlety.

---

31 For what follows, see Ronald Fagin and Joseph Halpern, 'Reasoning About Knowledge and Probability', *Journal of the Association for Computing Machinery*, 41 (1994), 340–367.

The second axiom (where i-probability sentences are sentences starting with $\mathbf{P}_i$ or Boolean combinations thereof) yields the players quite some measure of introspective power. It is easily seen, for instance, that the KnProb-axiom together with Cons-axiom and the necessitation rule entail positive as well as negative introspection.

The proof rules are thus:

MP If $\Sigma \vdash \varphi \rightarrow \psi$ and $\Sigma \vdash \varphi$, then $\Sigma \vdash \psi$.
Nec If $\vdash \varphi$, then $\vdash \Box_i \varphi$.
Ind If $\vdash \varphi \rightarrow \mathbf{E}_I(\varphi \wedge \psi)$, then $\vdash \varphi \rightarrow \mathbf{C}_I \psi$.

The first two rules of *modus ponens* and *necessitation* appear in the proof system of every epistemic logic. The last rule of *induction* is specific for proof systems that contain the E- and C-axiom.

To capture normal form game-playing situations we need four additional axioms.

Strat$_{\geq 1}$ $\bigvee_m \mathbf{i}_m$.
Strat$_{\leq 1}$ $\bigwedge_{m \neq n} \neg(\mathbf{i}_m \wedge \mathbf{1}_n)$.
KnStrat $\bigwedge_m (\Box_i \mathbf{i}_m \leftrightarrow \mathbf{i}_m)$.
KnUt $\mathbf{u}_i(k,l) = \mathbf{r} \rightarrow \Box_i \mathbf{u}_i(k,l) = \mathbf{r}$.

These axioms determine what players do, and what they know, when they play normal form games. The first axiom stipulates that every player plays at least one strategy, while the second axiom forbids any player to play more than one strategy. The KnStrat-axiom requires a player to have correct beliefs about what strategy he chooses. The KnUt-axiom requires players to have correct beliefs about their own utility functions.[32]

The precise formalisations of the game-theoretic solution concepts follow in the next two chapters. For the Nash equilibrium, no extra formal material is needed. For the iterated dominance solution concepts, I will develop a new way of axiomatisation.

1.2.2 A Logic for Extensive Games

Negation, connectives, and abbreviations are as before, as are, for the one-shot interpretation, the modal operators. For the many-moment interpretation doxastic or epistemic modalities $\Box_i^x$ are used to represent player i's beliefs or knowledge at the decision moment at which decision node x is reached, and super-scripted versions of $\mathbf{E}_I^x$, $\mathbf{C}_I^x$ and $\mathbf{P}_i^x(\cdot) = \cdot$ are defined similarly.

---

[32] Since any proof system contains the necessitation rule, players also believe (or know) these axioms to be true, believe them to believe them to be true, and so forth. This yields common beliefs about the possible actions, and about the fact that players know their utility. In the epistemic characterisation of mixed iterated strict weak dominance the last axiom takes a different form. See Section 2.4.

The roles decision nodes and decision moments play in extensive game-playing situations require me to restructure the original normal form language a little. On the basis of an arbitrary enumeration of all full strategies of some extensive form game, proposition letters $\mathbf{i}_k$ denote the kth such strategy; more precisely—the statement 'i plays her kth full strategy'. For decision node x (which is not necessarily a decision node where i has to move), proposition letter $\mathbf{i}_k^x$ states that player i chooses according to the kth strategy at all decision nodes in the subgame generated by x, and incidentally $\mathbf{i}_k(x)$ is used for the statement that, at her decision node x, player i chooses the action prescribed by her kth full strategy.[33] Utility statements need to be relativised as well. As before, the statement $\mathbf{u}_i(k,l) = \mathbf{r}$ captures the fact that 'the utility for player i, when full strategies k and l are being played, is r', and $\mathbf{u}_i^x(k,l) = \mathbf{r}$ restricts this statement to the subgame generated by x, meaning that 'the utility for player i, when the restrictions of full strategies k and l to the subgame generated by x are being played, is r'. Finally, a number of proposition letters are needed for the various principles of rationality such as $\mathbf{anrat}_i$, $\mathbf{nrat}_i$ and $\mathbf{rrat}$, to be explained later. Super-scripted, they express the respective relativised statements.

We need most of the earlier modal axioms in order to study extensive game-playing situations: the doxastic and epistemic axioms, the axioms for linear (in)equalities, the axioms for probability theory and the interrelation axioms, as well as the three proof rules. For one-shot analysis, we can adopt the earlier normal form version, but for the many-moment analysis we need to employ axioms and rules for such modalities as $\Box_i^x$, $\mathbf{P}_i^x(\cdot) = \cdot$, and so forth. It is plain how this is done consistently, though.

1.2.2.1 The One-Shot Interpretation

Two proof systems formalise the two interpretations of extensive game-playing situations. The one-shot interpretation has the following axioms.

Strat$_{\geq 1}$ $\bigvee_m \mathbf{i}_m$.

Strat$_{\leq 1}$ $\bigwedge_{m \neq n} \neg(\mathbf{i}_m \wedge \mathbf{i}_n)$.

KnStrat $\bigwedge_m (\Box_i \mathbf{i}_m \leftrightarrow \mathbf{i}_m)$.

KnUt $\mathbf{u}_i(k,l) = \mathbf{r} \rightarrow \Box_i \mathbf{u}_i(k,l) = \mathbf{r}$.

Sub$_1$ $\mathbf{i}_k^x \leftrightarrow \bigvee_{l \in D} \mathbf{i}_l$ where D contains those strategies coinciding with k on the subgame generated by x.

Sub$_2$ $\mathbf{i}_k(x) \leftrightarrow \bigvee_{l \in D} \mathbf{i}_l$ where D contains those strategies coinciding with k on decision node x.

UtSub $\mathbf{u}_i^x(k,m) = \mathbf{u}_i^x(l,n)$ whenever i's kth and lth, and j's mth and nth strategies coincide on the subgame generated by x.

KnUtSub $\mathbf{u}_i^x(k,l) = \mathbf{r} \rightarrow \Box_i \mathbf{u}_i^x(k,l) = \mathbf{r}$ for all decision nodes x.

---

[33] It is immaterial whether $\mathbf{i}_k^x$ and $\mathbf{i}_k(x)$ are really new proposition letters, or only abbreviations of the disjunction of the proposition letters of those strategies that coincide with the kth strategy on the subgame generated by x.

The first four axioms are copies of the conditions on normal form game-playing situations. Players pick exactly one full strategy, they know what they do, and they know what their utility functions are. Sub_1 states that the use of super-script is to talk about the restriction of some full strategy to the relevant subgame. Sub_2 ensures that function notation is present to report the action taken at some decision node. UtSub guarantees that the super-script works well when applied to utility functions. The KnUtSub-axiom, finally, is there to endow players with knowledge about their utility function in subgames.

1.2.2.2 The Many-Moment Interpretation

The following axioms fix the many-moment interpretation.

$\text{Strat}_{\geq 1}$ $\bigvee_m \mathbf{i}_m$.
$\text{Strat}_{\leq 1}$ $\bigwedge_{m \neq n} \neg(\mathbf{i}_m \wedge \mathbf{i}_n)$.
Sub_1 $\mathbf{i}^x_k \leftrightarrow \bigvee_{l \in D} \mathbf{i}_l$ where D contains those strategies coinciding with k on the subgame generated by x.
Sub_2 $\mathbf{i}_k(x) \leftrightarrow \bigvee_{l \in D} \mathbf{i}_l$ where D contains those strategies coinciding with k on decision node x.
UtSub $\mathbf{u}^x_i(k,m) = \mathbf{u}^x_i(l,n)$ whenever i's kth and lth, and j's mth and nth strategies coincide on the subgame generated by x.
KnStratM $\mathbf{i}_k \leftrightarrow \bigwedge_{p \preceq x} \Box^x_i \mathbf{i}_k(x)$.
KnUtM1 $\mathbf{u}_i(k,l) = \mathbf{r} \rightarrow \Box^x_i \mathbf{u}_i(k,l) = \mathbf{r}$ for all decision nodes x.
KnUtM2 $\mathbf{u}^y_i(k,l) = \mathbf{r} \rightarrow \Box^x_i \mathbf{u}^y_i(k,l) = \mathbf{r}$ for all decision nodes x and y.
KnWhere $\Box^x_i \bigwedge_j \bigvee_{y \prec x, j_k \in D} \mathbf{j}^y_k$ where D contains the proposition letters for those full strategies that are consistent with reaching x.

The first five axioms are those axioms from the one-shot interpretation that do not contain a $\Box_i$, and their motivation is similar. Of the last four axioms, KnStratM ensures that at every moment of a game-playing situation players know what they choose then and there.[34] The next two axioms ensure that players know their utilities in the entire game as well as in all subgames. The last axiom is there to guarantee that players know, at some decision moment, which decision node has been reached.

Finally, the solution concept of backward induction is defined in Chapter 3.

[34] This means that I adopt an *at choice*, rather than a *pre choice* conception of game-playing situations where the beliefs and preferences are considered just before the moment of choice, not at the moment of choice. See Wlodek Rabinowicz, 'Grappling with the Centipede: Defence of Backward Induction for BI-Terminating Games', *Economics and Philosophy*, 14 (1998), 115–119.

Part I
Epistemic Logic

Chapter 2
Normal Form Games

Epistemic characterisation theorems uncover the epistemic assumptions that underlie game-theoretic solution concepts. Their antecedents contain statements involving players' beliefs, preferences and rationality principles; their consequents describe actions. This, at least, is the ideal that the belief–desire outlook on human agency inspires, and this ideal is shared by most of the results from the Epistemic Programme.

Yet the Nash equilibrium already insults this ideal. The first of four solution concepts that I will examine in this chapter is characterised in two ways, one for actions in equilibrium, and one for—less intuitively intelligible—beliefs in equilibrium. A close reading of the logic reveals that the latter is the former with an epistemic modality prefixed, and that both suffer from irreparably illicit uses of the T-axiom entailing that beliefs, for the Nash equilibrium, have to be necessarily true, and that is conceptual nonsense.

I will subsequently deal with three iterated dominance concepts. Their characterisation results all fulfil the ideal, describing the solution concepts in endogenous terms without any inappropriate epistemic axioms. Iterated strict dominance uses plain rationality, but our axiomatisation of this notion is new. The Dekel–Fudenberg procedure (one round of elimination of weakly dominated strategies followed by infinitely many rounds of elimination of strictly dominated strategies) uses perfect rationality. The way I formalise it is similar to that of iterated strict dominance, but in order to analyse the plausibility of the rationality axioms, a comparison with Robert Stalnaker's game models will be informative. Finally, the third iterated dominance concept is a new one. It is arrived at when models are constructed of players whose information about each others' preferences is not completely accurate—*payoff-uncertainty*. Eddie Dekel and Drew Fudenberg investigated common knowledge of payoff-uncertainty and demonstrated that it leads to outcomes surviving the Dekel–Fudenberg procedure. If we use a different formalisation of payoff-uncertainty in which players have equiprobable beliefs about utility functions, we arrive at a concept called *mixed iterated strict weak dominance*.[1]

[1] The general form of the present chapter and the next has benefited from Pierpaolo Battigalli and Giacomo Bonanno, 'Recent Results on Belief, Knowledge and the Epistemic Foundations of Game

B. de Bruin, *Explaining Games*, Synthese Library 346,
DOI 10.1007/978-1-4020-9906-9_2, © Springer Science+Business Media B.V. 2010

2.1 The Nash Equilibrium

2.1.1 The Epistemic Characterisation Theorems

2.1.1.1 An Explicit Formalisation of Rationality

One epistemic characterisation theorem reveals that rationality, knowledge about utility, and knowledge about opponent strategies leads to a Nash equilibrium. Furthermore, if a player knows her opponent to be rational, knows that he knows the utility structure, and knows his probabilistic beliefs, these beliefs form a mixed strategy Nash equilibrium.[2] In order to understand the two results—a key claim here is that they do not conform to the general format of the Epistemic Programme—let me first define a number of concepts more precisely. A strategy profile $(a_1,\ldots,a_n)$ is a *Nash equilibrium* if and only if for all players i and all actions $b_i \in A_i$ it is true that

$$u_i(a_1,\ldots,a_i,\ldots,a_n) \geq u_i(a_1,\ldots,b_i,\ldots,a_n).$$

That is, no player can increase payoff by deviating unilaterally from the equilibrium.[3] In order to capture the Nash equilibrium in our formal framework, we do not need specific axioms, because utility comparisons can be effected merely by means of an immediate rendering of the principle of expected utility maximisation. Using the formalism introduced before, the axiom

MEU $\quad \mathbf{meu}_i \leftrightarrow \bigwedge_m((\Box_i \bigwedge_{k,l} \mathbf{u}_i(k,l) = \mathbf{r}_{i,k,l} \wedge \bigwedge_l \mathbf{P}_i(\mathbf{j}_l) = \mathbf{p}_l \wedge \mathbf{i}_m) \rightarrow$
$\bigwedge_k \sum_l \mathbf{p}_l \cdot \mathbf{r}_{i,m,l} \geq \sum_l \mathbf{p}_l \cdot \mathbf{r}_{i,k,l})$

states that player i is an expected utility maximiser whenever, if she decides to play her mth strategy in a situation in which she has certain beliefs about utility (captured by the $\mathbf{r}_{i,k,l}$) and about her prospective strategies (captured by the $\mathbf{p}_l$) then the mth strategy is better than any other, given her beliefs.[4] Around this axiom a proof

Theory', *Research in Economics*, 53 (1999), 149–225, and Adam Brandenburger, 'The Power of Paradox: Some Recent Developments in Interactive Epistemology', *International Journal of Game Theory*, 35 (2007), 465–492.

[2] Both theorems are due to Robert Aumann and Adam Brandenburger, 'Epistemic Conditions for Nash Equilibrium', *Econometrica*, 63 (1995), 1161–1180. See also Adam Brandenburger, 'Knowledge and Equilibrium in Games', *Journal of Economic Perspectives*, 6 (1992), 83–101. An earlier version is due to Wolfgang Spohn, 'How to Make Sense of Game Theory', in W. Balzer, W. Spohn and W. Stegmüller (eds.), *Studies in Contemporary Economics: Vol. 2: Philosophy of Economics* (Berlin: Springer, 1982), 239–270. Aumann and Brandenburger seem to have been unaware of Spohn's paper. Robert Stalnaker, 'On the Evaluation of Solution Concepts', *Theory and Decision*, 37 (1994), 49–73 gives a proof of the second characterisation theorem using methods from the model theory of modal epistemic logic.

[3] The conditions are sometimes referred to as the *Nash conditions*. The definition of Nash equilibrium is easily generalised so as to incorporate mixed actions as well.

[4] An unusual element in our formalisation is the appearance of a doxastic operator in front of the utility statement, deviating from the general set-up of explanations of human actions explored in Chapter 1. It is included here in order to make the study of the epistemic characterisation theorems

system $_{\Gamma}$**KTPmeu** is built comprising, besides MEU, the axioms Prop, Dual, K, T, the linear (in)equality axioms, the Kolmogorov axioms, the interrelation axioms, the proof rules modus ponens and necessitation, and the four axioms for normal form game-playing situations.

Table 2.1

Assumptions	
preferences	$\bigwedge_i \Box_i \bigwedge_{k,l} \mathbf{u}_i(k,l) = \mathbf{r}_{i,k,l}$
principle	$\bigwedge_i \mathbf{meu}_i$
beliefs	$\Box_2 \mathbf{1}_m \wedge \Box_1 \mathbf{2}_n$
Solution Concept	
player 1	$\bigwedge_k \mathbf{r}_{1,m,l} \geq \mathbf{r}_{1,k,l}$
player 2	$\bigwedge_l \mathbf{r}_{2,k,n} \geq \mathbf{r}_{2,k,l}$
Proof System	$_{\Gamma}$**KTPmeu**

Proof Two characterisation results are stated here, which reflect two ways to interpret mixed strategies. In the first interpretation, mixed strategies are objects of choice for the players, that is, players may deliberately use randomisation devices to generate probability measures on their pure strategies. In the second interpretation, by contrast, a mixed strategy of some player i does not so much constitute a strategy i can choose to perform, but represents the probabilistic beliefs i's opponents have about i's strategy choice. The first theorem gives a characterisation of the Nash equilibrium as an equilibrium of (pure or mixed) strategies. The second theorem deals with the Nash equilibrium of beliefs. The relevant ingredients of the theorems are listed in Tables 2.1 and 2.2.[5] Without loss of generality, I will restrict attention to a two-person normal form game.

Theorem 2.1 (Aumann and Brandenburger, 1995) *Let Γ be an N-person normal form game, and assume that the next three conditions hold.*

1. *All players know their own utility function.*
2. *All players are rational.*
3. *All players know each player's actual choice of action.*

Then the actual action profile played constitutes a Nash equilibrium.

more transparent, even though it could be omitted in the presence of the KnUt-axiom. Note further that whenever a decision or game theorist voices a claim to the effect that certain agents maximise their expected utility, he or she should be taken to view the agents as believing to be solving a maximisation problem corresponding to the beliefs and the utility functions that they think they have.

[5] The tables make fewer distinctions than later tables for characterisation results. This is in order to facilitate an easier comparison of the two versions of the epistemic characterisation of the Nash equilibrium.

It is shown that $\bigwedge_k \mathbf{r}_{1,m,l} \geq \mathbf{r}_{1,k,l}$ (the first clause of the condition for the solution concept). Inspection of the MEU-axiom shows that three elements are needed for the proof to work: the beliefs of player 1 about her opponent, the strategy she performs and her beliefs about the game structure.

The last clause is entirely trivial, so let me turn instead to the first two. By the rightmost conjunct of the condition about beliefs and the Cons-axiom for probability, it follows that $\mathbf{P}_1(\mathbf{2}_n) = 1$. On the basis of Strat$_{\leq 1}$ it can be proven that $\bigwedge_{l \neq n} \Box_1(\neg \mathbf{2}_l)$ which—using $\mathbf{P}_i(\varphi) = 1 - \mathbf{P}_i(\neg\varphi)$ and the Cons-axiom—is equivalent to $\bigwedge_{l \neq n} \mathbf{P}_1(\mathbf{2}_l) = 0$. In other words, the assumptions imply that all except one of the $\mathbf{p}_{i,l}$ vanish. This establishes the clause about the beliefs player 1 has about her opponent. The second clause about the performed strategy follows easily from the assumed condition about beliefs together with the T-axiom for the $\Box_2$; that is, they yield $\mathbf{1}_m$, player 1 plays her mth strategy.

These ingredients, and the assumption that player 1 is rational, make it easy to prove $\bigwedge_k \sum_l \mathbf{p}_{i,l} \cdot \mathbf{r}_{1,m,l} \geq \sum_l \mathbf{p}_{i,l} \cdot \mathbf{r}_{1,k,l}$. This is equivalent to

$$\bigwedge_k \mathbf{r}_{1,m,l} \geq \mathbf{r}_{1,k,l},$$

because, as I have shown above, all except one of the $\mathbf{p}_{i,l}$ are zero.

Table 2.2

Assumptions	
preferences	$\mathbf{EE}(\bigwedge_i \bigwedge_{k,l} \mathbf{u}_i(k,l) = \mathbf{r}_{i,k,l})$
principles	$\bigwedge_i \Box_i \mathbf{meu}_j$
performed action	$\Box_1 \bigwedge_m \mathbf{P}_2(\mathbf{1}_m) = \mathbf{p}_m$
	$\Box_2 \bigwedge_n \mathbf{P}_1(\mathbf{2}_n) = \mathbf{q}_n$
Solution Concept	
player 1	$\bigwedge_m(\mathbf{q}_m \neq 0 \rightarrow \bigwedge_l \sum_k \mathbf{p}_k \cdot \mathbf{r}_{2,k,m} \geq \sum_k \mathbf{p}_k \cdot \mathbf{r}_{2,k,l})$
player 2	$\bigwedge_n(\mathbf{p}_n \neq 0 \rightarrow \bigwedge_k \sum_l \mathbf{q}_l \cdot \mathbf{r}_{1,n,l} \geq \sum_l \mathbf{q}_l \cdot \mathbf{r}_{1,k,l})$
Proof System	$_\Gamma$**KTPmeu**

I will now turn to the second characterisation result. Here $\Delta(A_j)$ denotes the set of probability measures on A_j, the set of j's pure strategies. While the previous result holds for normal form games of arbitrarily many players, this is not true here. Table 2.2 lists the essential ingredients of the theorem.

Theorem 2.2 (Aumann and Brandenburger, 1995) *Let Γ be a two-person normal form game, and assume that the next three conditions are satisfied.*

1. *Each player knows that any player knows the utility functions of both players.*
2. *Each player knows that her opponent is rational.*

3. *Each player i knows that her opponent j has probabilistic beliefs about i's actions given by some $\sigma_j \in \Delta(A_i)$.*

Then (σ_2, σ_1) is a mixed Nash equilibrium.

The formalisation of the assumptions underlying the theorem is undemanding.[6] In order to explain the idea behind the conclusion, a precise definition of what it means for a strategy to be a best response has to be given. If player i has probability measure P_i on her opponents' actions, the expected utility of playing some action $a_i \in A_i$ is defined as

$$\mathrm{EU}_i(\mathsf{P}_i, a_i) = \sum_{a_{-i} \in \prod_{j \neq i} A_j} \mathsf{P}_i(a_{-i}) u(a_i, a_{-i}).$$

A pure strategy a_i is a best response among A_i to some probability measure P_i on $\prod_{j \neq i} A_j$ whenever for any action $b_i \in A_i$ the expected utility of playing b_i is at most as high as the expected utility of playing a_i, or formally,

$$\sum_{a_{-i} \in \prod_{j \neq i} A_j} \mathsf{P}_i(a_{-i}) u(a_i, a_{-i}) \geq \sum_{a_{-i} \in \prod_{j \neq i} A_j} \mathsf{P}_i(a_{-i}) u(b_i, a_{-i}).$$

A useful lemma reveals that on the basis of the assumptions made by the theorem it is sufficient to show that any strategy receiving non-zero probability by $\mathbf{P}_i$ (a strategy in the *support* of $\mathbf{P}_i$) is a best response by player i against $\mathbf{P}_{-i}$.[7]

Lemma 2.1 *Let $\Gamma = (I, (A_i)_i, (u_i)_i)$ be an N-person normal form game with finite strategy sets. Then a profile of mixed actions $(\sigma_1, \dots, \sigma_N)$ is a Nash equilibrium if and only if for all players i each strategy in the support of σ_i is a best response to σ_{-i}.*

Given the way the $\mathbf{p}_k$ and the $\mathbf{q}_l$ are fixed in the assumptions of the second characterisation result, this is exactly what the conditions for players 1 and 2 mean.

Proof Proving the condition for player 1, assume that for an arbitrary m we have $\mathbf{q}_m \neq 0$. From the condition about principles and some modal reasoning on the MEU-axiom it follows that

$$\Box_1((\Box_2 \bigwedge_{k,l} \mathbf{u}_2(k,l) = \mathbf{r}_{2,k,l} \wedge \bigwedge_k \mathbf{P}_2(\mathbf{1}_k) = \mathbf{p}_k \wedge \mathbf{2}_m) \rightarrow$$
$$\bigwedge_l \sum_k \mathbf{p}_k \cdot \mathbf{r}_{2,k,m} \geq \sum_k \mathbf{p}_k \cdot \mathbf{r}_{2,k,l})$$

Let me note this as $\Box_1(\varphi_m \rightarrow \psi_m)$. It has to be shown that ψ_m; and as this is a linear inequality not changing its truth value in the scope of modal operators, it suffices to

[6] In games containing no pure Nash equilibria, no strategies 1_k and 2_l that would satisfy the assumptions can be found. Either 1_k will not be a best response to 2_l or the converse. Modelling the mixed extension of the game (containing all continuum many mixed strategies) is a way out.

[7] Martin Osborne and Ariel Rubinstein, *A Course in Game Theory* (Cambridge: MIT Press, 1994), 51.

prove $\Diamond_1 \psi_m$. Standard modal reasoning shows that, since we have $\Box_1(\varphi_m \rightarrow \psi_m)$, it suffices to prove $\Diamond_1 \varphi_m$.

It follows directly from the T-axiom and the assumptions on preferences and performed actions (first line) that $\Diamond_1 \Box_2 \bigwedge_{k,l} \mathbf{u}_2(k,l) = \mathbf{r}_{2,k,l}$ and $\Diamond_1 \bigwedge_k \mathbf{P}_2(\mathbf{1}_k) = \mathbf{p}_k$. Since $\mathbf{q}_m \neq 0$, we find $\mathbf{P}_1(\mathbf{2}_m) \neq 0$ from the assumption on performed action (second line) plus the T-axiom, and with the help of the Cons-axiom and some elementary Kolmogorov reasoning it follows that $\Diamond_1 \mathbf{2}_m$. Together this establishes

$$\Diamond_1(\Box_2 \bigwedge_{k,l} \mathbf{u}_2(k,l) = \mathbf{r}_{2,k,l} \wedge \bigwedge_k \mathbf{P}_2(\mathbf{1}_k) = \mathbf{p}_k \wedge \mathbf{2}_m),$$

concluding the proof.

2.1.2 Discussion

Chapter 1 explained the general logical form epistemic characterisation results have to take if they are to conform to the belief–desire framework which explains actions in terms of reasons. Antecedent conditions contain statements about players' preferences, beliefs and rationality principles; consequent conditions describe the actions the players perform; and together they reveal the reasons or motivations players have to perform actions satisfying certain solution concepts. Yet the two epistemic characterisation results discussed here deviate from this set-up. Neither one of them contains action statements in the consequent, and the second theorem's antecedent does not contain any statement to the effect that the players are rational.

Two factors are responsible for this unfortunate situation, or so I argue here. First, that no actions are referred to in the consequent arises from the adoption of the T-axiom. But removing the T-axiom, while making the theorems conform to the belief–desire format in some respects, would require beliefs to be necessarily true, which is a very dubious thing to require in explanations of actions. Second, by examining possible equilibration processes the rationality in the second theorem turns out to be the *theoretic* rationality of belief adoption and revision, not the *practical* rationality of agency. But no such rationality is referred to in the second epistemic characterisation theorem.

While both observations already cast serious doubt on the solution concept of Nash—not on the epistemic characterisation results as such, of course—a final blow is dealt when I examine how the Nash equilibrium fares under the ban on exogenous information.

2.1.2.1 The Axiom of Truth

To start, I should note that while it is easy to circumvent using the proof system of $_{\Gamma}$**KTPmeu** by adding to the assumptions the relevant clauses ($\Box_i\varphi$ rewrites as

$\Box_i\varphi \wedge \varphi$, and so on) this is undesirable from a conceptual point of view, as it would contradict the idea that epistemic characterisation results lay bare the players reasons for performing certain actions, or in the second theorem, adopting certain beliefs. If, for instance, to the assumptions of the first theorem a statement was added to the effect that $\mathbf{1}_k \wedge \mathbf{2}_l$, a genuine motivation for the players' actions would no longer be expressed in the antecedent of the characterisation result. A motivation to perform some action a must not contain a statement that a is being performed. Rather, it must give reasons for why it is. Similarly, for the second theorem, if it is going to specify the agents' motivations for adopting certain beliefs, then the beliefs must not be part of the motivations. Once the T-axiom is abandoned, both characterisation results become altogether meaningless from the belief–desire point of view—antecedents should state motivations.

Yet the desired form can be obtained no better in a context in which the T-axiom features. The T-axiom requires that the beliefs of the players be necessarily true. But for an agent, the sole fact that a belief is true plays no motivational role whatsoever. A player may believe that by performing some action she reaches some desired outcome, and this belief may be true.[8] But the fact that a belief is true does not figure in her motivation to perform the action. The belief suffices; truth is unnecessary.[9]

	2_1	2_2
1_1	(1,-1)	(-1,1)
1_2	(-1,1)	(1,-1)

Fig. 2.1

It might be thought that using the T-axiom in the epistemic characterisation of the Nash equilibrium just reflects a philosophical disagreement with the view of truth in action explanations, but if the present analysis is correct, veridicality in the characterisation results figures for simpler reasons: without it, they would not work. Without the T-axiom, it would be provable that player 1's strategy is a best response to her (in this case possibly false) beliefs about 2, but that is all. To see this, consider the game of Matching Pennies shown in Figure 2.1. If the assumptions that underlie the first characterisation result are satisfied with $k = l = 1$, proof system $_{\Gamma}$**KPmeu** (note the absence of the T-axiom) allows me to prove that $\mathbf{1}_1 \wedge \mathbf{2}_2$. This is a unique pair of actions, but it is not a Nash equilibrium. Without the T-axiom,

[8] Many beliefs in subsequent epistemic characterisation results will likewise be contingently true.

[9] If knowledge possesses causal efficacy (see note 25), this is merely due to the fact that knowledge is more than true belief. The causal efficacy is not due to the truth of knowledge.

agents cannot be guaranteed to end up with unique probabilistic beliefs that support Nash equilibria.[10]

2.1.2.2 The Rational Equilibration of Beliefs

In order to obtain a clearer view of what it means, in the second theorem, for beliefs to be in equilibrium, let me first take a brief look at the two main views of what probability measures (that is, mixed strategies) actually model. According to the most common conception, mixed strategies involve real objects of choice, modelling players who deliberately choose to randomise over pure strategies. As Lemma 2.1 reveals, however, a peculiar feature of mixed strategies is that players are indifferent among all pure actions that receive non-zero probability in the mixture, making it a little difficult to explain why players would actually randomise instead of choosing pure strategies. A second interpretation conceives of player i's mixed strategies as if they were the beliefs of other players about i, and although calling them *strategies* is rather confusing, mathematically there is nothing wrong with this interpretation, because player i's mixed strategies are probability measures on her actions.

While the first theorem is readily consistent with the interpretation of mixed strategies as objects of choice, the second interpretation of mixtures as beliefs is the natural environment of the second characterisation theorem.[11] Yet to understand how that interpretation can provide such a context, I will reconsider the logical form of the results. As I have suggested, from the viewpoint of a belief–desire approach to action explanation, the first characterisation result attempts to specify the motivations that a player has to choose a particular action. Nothing precludes adopting the same view of the second theorem and interpreting it as a specification of the motivations players have to form particular beliefs. The antecedent conditions state such things as that player 1 knows that 2 knows her payoffs, that 1 knows that 2 is rational, and that 1 knows 2's probabilistic beliefs about 1, while the consequent suggests as a criterion for belief formation that if player 1's probability measure assigns positive probability to some action 2_l, this action is a best response, for player 2, to the probabilistic beliefs player 1 knows 2 to have. This is reasonable, because otherwise a contradiction with player 1's knowledge of 2's rationality would be immediate.

In order to evaluate the general plausibility of such a belief formation process, let me take a slightly more abstract look at the difference between the two characteri-

[10] With respective differences taken into consideration this is true for the second theorem as well, problematising the motivation of the beliefs rather than that of the actions. The impossibility does not depend on the fact that Matching Pennies contains no pure Nash equilibria. If both players get an additional third strategy yielding both zero against every strategy, the strategy profile consisting of both players' third strategies constitutes a pure Nash equilibrium. From the above assumptions it is impossible to prove that this equilibrium will be met.

[11] Adopting an infinitary language is necessary to incorporate mixed strategies in my framework. Aumann and Brandenburger, art. cit. 1162 agree, writing that the first theorem 'applies to pure strategies' and 'also to mixed actions, under the traditional view of mixtures as conscious randomisations', but that 'the context of our main results [the second theorem]' is formed by the second interpretation.

Table 2.3

	Actions	*Beliefs*
Assumptions		
preferences	$\Box_i\mathbf{u}_i$	$\Box_j\Box_i\mathbf{u}_i$
principles	$\mathbf{meu}_i$	$\Box_j\mathbf{meu}_i$
performed action	$\Box_i\mathbf{j}_n$	$\Box_j\Box_i\mathbf{j}_m$
Solution Concept	Nash($\mathbf{i}_m,\mathbf{j}_n$)	Nash($\Box_j\mathbf{i}_m,\Box_i\mathbf{j}_n$)
Proof System	$_\Gamma$**KTPmeu**	$_\Gamma$**KTPmeu**

sation results, substituting the probabilistic beliefs $\mathbf{P}_i$ from the second theorem with non-probabilistic beliefs $\Box_i$.[12] This transformation reveals that the second theorem can be obtained from the first by means of simply prefixing every clause with an epistemic modality (see Table 2.3) and this is conceptually very intriguing, because it shows that the argumentation used in the first theorem is used in the second—with a prefixed epistemic operator, of course. If the first theorem attempts to expose player *i*'s motivations which underlie some choice of strategy, the second theorem studies player *j* reflecting on *i*'s motivational reasoning, and both set out to accomplish this in a setting of what may be called *equilibration.*

It is not difficult to understand a reasonable equilibration process motivating the first theorem, showing convergence to a situation in which players are rational, know their own utility, and know their opponents' prospective actions. The respective actions players 1 and 2 perform in the equilibrium, for instance, keep each other in equilibrium under the assumption that, first, if player 2 chooses to perform another action, player 1 perceives this in some way; second, if player 2 chooses to perform another action, player 1 will respond by switching to another strategy rationally; and third, player 1 knows her payoff structure. This, one could argue, is the content of the characterisation result. The first condition (perception) corresponds to $\Box_i\mathbf{j}_n$; the second (rationality), to $\mathbf{meu}_i$; and the third (payoff) to $\Box_i\mathbf{u}_i$.

A similar equilibration process is much less plausible for the second theorem. It would require that, first, if player 2 adopts a different belief about 1, player 1 perceives this change; second, if player 2 assumes another belief about 1, player 1 responds by changing her own beliefs in a theoretically rational way; and third, to ensure that some epistemic condition on utility functions holds. If, for instance, player 1 perceives a belief change in 2, then it is theoretically rational for 1 to change her beliefs about 2 as well, because, in fact, if player 2 changes his beliefs about 1, he

[12] Only the third condition changes in this substitution process, for instead of requiring that 1 know the probabilistic beliefs $\mathbf{P}_2$ player 2 possesses about 1's actions, it is now required that player 1 know player 2's full beliefs $\Box_2$ about 1's actions. For some $\mathbf{1}_m$ we have $\Box_1\Box_2\mathbf{1}_m$ and for some best response $\mathbf{2}_n$ to $\mathbf{1}_m$, it follows that $\Box_1\mathbf{2}_n$. Given player 1's beliefs as specified in the assumptions of the theorem, no other beliefs would be reasonably motivated.

probably changes his choice of action.[13] In iterated game-playing situations, players generally perceive their opponents' choices from the previous rounds, which may lead them to readjust their own strategy choice in the next round and this is a pretty straightforward way to think of equilibration.[14] But while it would be going too far to claim that beliefs never change perceptibly, it is a far less common phenomenon than perception of strategy choice. To the extent that sense can be made of such an equilibration process for beliefs, beliefs in equilibrium are meaningful. As long as it remains unspecified what rational belief adoption means, however, the second theorem contradicts the general form of epistemic characterisation theorems as much as the first.[15]

2.1.2.3 The Ban on Exogenous Information

To summarise this discussion, it has become clear that neither of the two theorems conforms to the format for epistemic characterisation results laid out in Chapter 1, as each requires the T-axiom, since neither provides for action statements in the consequent of the implication, and because the second does not refer to rationality in its antecedent. Abstracting from probabilistic reasoning, I have argued that the first theorem transforms into the second if an epistemic modality is prefixed to the assumptions underlying the first, and this led to a discussion about beliefs in equilibrium—meaningful only if, rather implausibly, belief changes are perceptible.

Reinforcing this critique, I will examine how the Nash equilibrium fares under the ban on exogenous information. At first glance it might seem that no statement about the origin of the beliefs of the players comes with the epistemic characterisation of the Nash equilibrium; it looks entirely neutral as to whether the players have formed their beliefs on the basis of exogenous or endogenous reasoning. But if this were the case, purely endogenously formed beliefs could support the Nash equilibrium, and an alternative epistemic characterisation result would need to be developed. As Aumann and Brandenburger observe, however, typically endogenous concepts involving (variants of) common belief—long held to support the

[13] This example works provided that player 2 maximises expected utility and knows his payoff, which shows, more generally, that the equilibration process assumes that players believe each other to be practically rational and to know their utility functions. Common belief about practical rationality and utility does not seem necessary, though, because the equilibration process involves rational responses to perceived changes, not to (hypothetical) changes derived from general assumptions involving common beliefs.

[14] Witness, e.g., John Nash 'Non-Cooperative Games', Ph.D. diss. (Princeton University, 1950), 21–26. Also see Larry Samuelson, *Evolutionary Games and Equilibrium Selection* (Cambridge, Mass.: MIT Press, 1997).

[15] Hans Jørgen Jacobsen, 'On the Foundations of Nash Equilibrium', *Economics and Philosophy*, 12 (1996), 67–88 provides an alternative justification for the beliefs reading of the Nash equilibrium. His argument depends on assumptions about the concept of solution concept that I criticise in Chapter 5.

Nash equilibrium—are completely ineffective, and there is consequently no hope for an endogenous approach to this solution concept.[16]

This adds to the unacceptability of the T-axiom in unexpected ways. While this axiom is, as I have shown, already problematic for reasons regarding the logic of action explanation, it is completely inappropriate in an exogenous setting for the additional reason that exogenously formed beliefs based on statistics and the like are never true by necessity. Certain belief formation practices may be more accurate than others and more theoretically rational, but they can never be so good that they produce true beliefs only—by necessity, so to speak. The epistemic characterisation of the Nash equilibrium not only shows that the concept lacks any endogenous backing; it also makes clear that it cannot be justified in exogenous terms.

2.2 Iterated Strict Dominance

2.2.1 The Epistemic Characterisation Theorem

Common belief about rationality and utility entails iterated strict dominance.[17] In order to grasp this result formally, I will develop an implicit, inductive axiom system to capture rationality rather than the more usual explicit rendering of rationality as expected utility maximisation which I considered in the characterisation of the Nash equilibrium. To set the stage, a pure strategy $a_i \in A_i$ of player i is strictly dominated by a mixed strategy σ_i against all pure strategies of his opponents in $\prod_{j \neq i} A_j$ if and only if

$$u_i(\sigma_i, a_{-i}) > u_i(a_i, a_{-i})$$

for all i-deleted action profiles $a_{-i} \in \prod_{j \neq i} A_j$, and a strategy is *strictly dominated* in a game whenever there is a mixed strategy that strictly dominates it. If from a normal form game Γ all actions are removed not in sets X_i and $\prod_{j \neq i} X_j$, leaving the utility function intact, the resulting game is the game *spanned* by those sets, and a strategy is called strictly dominated in a game spanned by sets X_i and $\prod_{j \neq i} X_j$ whenever

---

[16] See, e.g., ibid., 1162.

[17] B. Douglas Bernheim, 'Rationalizable Strategic Behavior', *Econometrica*, 52 (1984), 1007–1028, David Pearce, 'Rationalizable Strategic Behavior and the Problem of Perfection', *Econometrica*, 52 (1984), 1029–1050, and Wolfgang Spohn, 'How to Make Sense of Game Theory' (1982) were the first to take an epistemic look at iterated strict dominance. Bernheim and Pearce developed their ideas independently and published them in tandem in the same issue of *Econometrica*. Neither of them seems to have been aware of Spohn's work. Tommy Tan and Sergio Werlang, 'The Bayesian Foundations of Solution Concepts of Games', *Journal of Economic Theory*, 45 (1988), 370–391 gave the first formal treatment in terms of type spaces using methods from probability and measure theory. Robert Stalnaker, 'On the Evaluation of Solution Concepts', *Theory and Decision*, 37 (1994), 49–73 proves the result in the a model theory for epistemic logic. Common belief was introduced in the context of games by Morris Friedell, 'On the Structure of Shared Awareness', *Behavioral Science*, 14 (1969), 28–39 and David Lewis, *Convention* (Cambridge, Mass.: Harvard University Press, 1969).

there is a mixture over elements from X_i satisfying the above strict inequality for all i-deleted strategy profiles in $\prod_{j \neq i} X_j$. Notation

$$\text{nsd}_i(X_i, X_{-i})$$

is used to denote the set of those pure strategies that are not strictly dominated, or strictly *undominated*, in the game spanned by X_i and $\prod_{j \neq i} X_j$, usually omitting explicit reference to the underlying game. With slight abuse of notation nsd_i and its variants are used for functions with different domains.

Strict dominance forms the basis of iterated strict dominance. First, define the set of pure strategies of player i as $\mathsf{S}_i^0 = A_i$, and the set of her mixed strategies as $\Sigma_i^0 = \Delta(A_i)$. Then collect recursively in set S_i^{n+1} those elements from S_i^n for which there is no $\sigma_i \in \Sigma_i^n$ such that

$$u_i(\sigma_i, a_{-i}) > u_i(a_i, a_{-i})$$

for all $a_{-i} \in \prod_{j \neq i} \mathsf{S}_j^n$, as well as in set Σ_i^{n+1} those elements σ_i from $\Delta(A_i)$ for which $\sigma_i(a_i) > 0$ only if $a_i \in \mathsf{S}_i^n$. Then, set

$$\mathsf{S}_i^\infty = \bigcap_n \mathsf{S}_i^n$$

and collect finally in Σ_i^∞ all elements σ_i from $\Delta(A_i)$ such that there is no mixed strategy σ_i' such that for all $a_{-i} \in \prod_{j \neq i} \mathsf{S}_j^\infty$ it holds that

$$u_i(\sigma_i', a_{-i}) > u_i(\sigma_i, a_{-i}).$$

These two sets determine the solution concept of iterated strict dominance. The set S_i^∞ contains the pure actions of player i that survive the iterated elimination of strictly dominated strategies. The set Σ^∞ is the set of mixed strategies surviving iterated strict dominance.[18] Slightly abusing the earlier notation,

$$\mathsf{S}_i^{n+1} = \text{nsd}_i(\mathsf{S}_j^n, \mathsf{S}_{-i}^n)$$

defines equally well.

2.2.1.1 An Implicit, Inductive Formalisation of Rationality

In my discussion of the logical form of epistemic charcacterisation results it became clear that three key elements in the antecedent are supposed to entail information about the strategies the players choose: beliefs, preferences and rationality. If you know what a player believes about her opponents, if you know what her utility

---

[18] In general Σ_i^∞ may be smaller than the set $\Delta(\mathsf{S}_i^\infty)$, as there may be mixed strategies over S_i^∞ that are dominated. See, e.g., Drew Fudenberg and Jean Tirole, *Game Theory* (Cambridge, Mass.: MIT Press, 1991), 45–46.

function looks like, and if, in addition, you know that she is an expected utility maximiser, then you know how she will act.

The proposed formalisation of rationality exploits the fact that sometimes not all three elements are needed to obtain (some) information about what a player will choose. Even in the absence of information about a player's beliefs, you can still exclude her from playing strategies that are bad no matter what her opponents do—as long as you know her utility function and that she is rational. However limited such information may be, it is nonetheless fruitful to consider this basic step of reasoning in isolation. Formally, using $\mathbf{rat}_i$ for rationality, the axiom

$$\text{Rat}_{bas} \quad \mathbf{rat}_i \rightarrow \text{nsd}_i(A_i, A_j)$$

states that if player i with a utility function as in the relevant underlying game is rational, she chooses a strategy that is not strictly dominated in the game spanned by A_i and A_j (the original game). This axiom gives an implicit definition of rationality in terms of the exclusion of strictly dominated strategies, and I will turn to different axioms characterising alternative concepts of rationality shortly. But the main idea now is that the Rat_{bas}-axiom captures the base case where, in complete absence of information about a player's beliefs, you are still able to exclude her from playing certain strategies—the really bad ones.

With information about a player's beliefs, you can say much more of course, and this is the content of the inductive axiom

$$\text{Rat}_{ind} \quad (\mathbf{rat}_i \wedge \Box_i X_i \wedge \Box_i X_j) \rightarrow \text{nsd}_i(X_i, X_j).$$

Suppose that player i (with preferences as in the game matrix) acts on the principle of rationality to exclude strictly dominated strategies. If you have information to the effect that (for whatever reasons) she believes that she will choose a strategy from set X_i and that her opponent j will choose a strategy from set X_j, then you may conclude that she chooses a strategy that is not strictly dominated in the game spanned by X_i and X_j, or so the axiom says. The inductive axiom is an axiom schema for two players $i \neq j$ for all subsets $X_i \subseteq A_i$ and $X_j \subseteq A_j$ of the relevant game. This is easily generalised to game-playing situations with more than two players.

To prove the epistemic characterisation theorem of iterated strict dominance to the effect that it results from common knowledge of rationality and utility, a proof system ${}_\Gamma\mathbf{K_{EC}rat}$ is used consisting of Prop, Dual, K, E, C, the proof rules modus ponens, necessitation and induction, and the four axioms for normal form game-playing situations, as well as the basis and inductive rationality axioms. The assumptions of the characterisation result are listed in Table 2.4.

Theorem 2.3 (Bernheim and Pearce, 1984) *Let Γ be a two-person normal form game. Assume that the following two conditions are true.*

1. *There is common true belief among the players about the utility functions of all players.*
2. *It is common true belief among the players that they are rational.*

Then the players play strategies that survive the iterated elimination of strictly dominated strategies.

Table 2.4

Assumptions	
preferences	$\bigwedge_{i,k,l} \mathbf{u}_i(k,l) = \mathbf{r}_{i,k,l}$
principles	$\bigwedge_i \mathbf{rat}_i$
beliefs	
preferences	$\mathbf{C} \bigwedge_{i,k,l} \mathbf{u}_i(k,l) = \mathbf{r}_{i,k,l}$
principles	$\mathbf{C} \bigwedge_i \mathbf{rat}_i$
performed action	–
Solution Concept	
player i	S_i^∞
Proof System	$_\Gamma\mathbf{K}_{\mathbf{EC}}\mathbf{rat}$

The proof of the theorem ignores the assumptions about the preferences of the players because firstly, it does adds little to the conceptual and technical understanding of the characterisation result to spell out common knowledge of utility structure—it is trivial here—and secondly, because a non-trivial study of knowledge about utility is examined in the investigation of a new concept of mixed iterated strict weak dominance in Section 2.4.

Proof Abbreviate $\bigwedge_i \mathbf{rat}_i$ by **rat**, recall that $\mathbf{E}^{\leq n}\varphi$ abbreviates $\mathbf{E}\varphi \wedge \mathbf{E}^2\varphi \wedge \cdots \wedge \mathbf{E}^n\varphi$ (suppressing the subscript), and note that it is not the case that $\vdash_{_\Gamma\mathbf{K}_{\mathbf{EC}}\mathbf{rat}} \mathbf{E}^{\leq n}\mathbf{rat}_i \to \mathbf{rat}_i$. I show by induction on n that

$$\forall n \vdash (\mathbf{rat} \wedge \mathbf{E}^{\leq n}\mathbf{rat}) \to \bigwedge_j S_j^{n+1},$$

where the set of proposition letters for iteratively strictly undominated strategies of player i is denoted S_i^∞, and super-scripts n indicate what remains after n rounds of elimination. The case of $n = 0$ is the case of the Rat_{bas}-axiom. As inductive hypothesis have a proof of

$$(\mathbf{rat} \wedge \mathbf{E}^{\leq n}\mathbf{rat}) \to \bigwedge_j S_j^{n+1}.$$

By the (derivable) necessitation rule for **E** the statement

$$\mathbf{E}((\mathbf{rat} \wedge \mathbf{E}^{\leq n}\mathbf{rat}) \to \bigwedge_j S_j^{n+1})$$

is then proved as well, and the (derivable) K-axiom for **E** then yields

$$\mathbf{E}^{\leq n+1}\mathbf{rat} \to \mathbf{E}\bigwedge_j S_j^{n+1}.$$

The Rat_{ind}-axiom for i states that

$$(\mathbf{rat}_i \wedge \Box_i S_i^{n+1} \wedge \Box_i S_j^{n+1}) \rightarrow \mathrm{nsd}_i(S_i^{n+1}, S_j^{n+1}),$$

of which the consequent is the same as S_i^{n+2} by definition. Since

$$\mathbf{E}\bigwedge_j S_j^{n+1} \rightarrow \bigwedge_k \Box_k \bigwedge_j S_j^{n+1}$$

it can be concluded that

$$(\mathbf{rat} \wedge \mathbf{E}^{\leq n+1}\mathbf{rat}) \rightarrow \bigwedge_j S_j^{n+2}.$$

This concludes the induction on n. Introducing the **C**-operator, this establishes

$$\forall n \vdash_{\Gamma \mathbf{K_{EC}}\mathbf{rat}} (\mathbf{rat} \wedge \mathbf{Crat}) \rightarrow \bigwedge_j S_j^{n+2}.$$

The S_i^n, including pure actions only, are of the form $\{\mathbf{i}_k \mid k \in I_n\}$ for non-empty finite index sets satisfying $I_n \supseteq I_{n+1} \supseteq \cdots \supseteq I_{n0} = I_{n0+1} = \cdots = \bigcap_n I_n \neq \emptyset$. That means that in the limit it holds that $S_i^\infty = \{\varphi_k \mid k \in I_{n0}\} = \bigcap_n S_i^n$. From the previous provability statement it follows that

$$\forall n \vdash_{\Gamma \mathbf{K_{EC}}\mathbf{rat}} (\mathbf{rat} \wedge \mathbf{Crat}) \rightarrow \bigwedge_j \bigcap_n S_j^n,$$

which is equivalent to what had to be shown.

2.2.2 *Discussion*

2.2.2.1 Axioms T and K, and the Rule of Necessitation

I have shown that the epistemic characterisation of the Nash equilibrium does not conform to the general format of action explanations employed in the Epistemic Programme : apart from not stating any actions in the consequent, the characterisation's use of the T-axiom contradicts the condition that beliefs are not necessarily true; with the T-axiom, however, the result just does not work.

The present theorem, by contrast, seems to conform. It derives actions surviving the iterated elimination of strictly dominated strategies from assumptions on the players' preferences, rationality principles and beliefs. In order to show truly that it does indeed follow the format, I have to show that nowhere are necessarily true beliefs needed, for while the other conspicuous requirements are obviously met, the mention of common *true* belief in the theorem may elicit suspicion.

This is rather easy, however. Sure, the result could be represented as

$$\vdash_{\Gamma\mathbf{KT_{EC}rat}} \mathbf{Crat} \rightarrow \bigwedge_i S_i^{\infty}.$$

But since it is different from the characterisation of the Nash equilibrium, it need not, because if we remove the T-axiom and add the condition $\bigwedge_i \mathbf{rat}_i$ to the antecedent, not only do we obtain an equally valid statement, but we also (and more importantly) get a statement that is in complete agreement with the view of action explanation favoured by the Epistemic Programme. There are beliefs, there are desires (implicit, here, in the utility functions) and there is rationality. That some of the beliefs concern rationality and that they are true, however, is not a necessity. The antecedent condition describes a situation where players form correct beliefs about rationality contingently, not necessarily. Thus the T-axiom could be used, but need not be used. In other words, it is not hidden in the theorem, and this means that the players are not committed to veridical beliefs.

	2_1	2_2	2_3
1_1	(1,5)	(0,1)	(1,3)
1_2	(5,0)	(0,0)	(3,1)
1_3	(3,3)	(1,1)	(2,2)

Fig. 2.2

Let me turn to another axiom. While in the proof of this and other characterisation results the K-axiom rarely appears, this axiom cannot be missed, and it is worthwhile to highlight its function. Suppose, for the sake of argument, that the antecedent assumptions of the epistemic characterisation theorem of iterated strict dominance hold in a game-playing situation of the game shown in Figure 2.2. In that case, player 1 believes that 2 is rational in the sense that he excludes strictly dominated strategies, she believes that players 2's preferences are as in the game matrix, and she believes the implication that if any player is rational and possesses these preferences, then he or she will never play his or her second strategy. Player 1, that is, has a belief about rationality, a belief about preference, and she has a belief about an implication concerning their connection.

In order to make a choice, however, player 1 needs a belief about player 2's prospective action, and she derives information to that effect from the beliefs stipulated in the previous paragraph by applying the K-axiom. Without this axiom, she would not be in the position to conclude that player 2 does not play his second strategy.

A similar argument shows that the necessitation rule is essential. The beliefs that play a part in epistemic characterisation theorems are of two kinds. Some concern contingencies such as whether your opponent is rational, or which utility function he has. Other beliefs, by contrast, are tautologies in the sense that they follow automatically from other beliefs you hold. Suppose, for instance, that you believe that your opponent maximises expected utility and has a utility function such that his first strategy is strictly dominated. These beliefs surely concern contingencies. But you may also hold the belief that players who maximise expected utility avoid strictly dominated strategies, and this belief is tautologous in that it merely reflects the logical consequences of certain propositions. The above analysis of the role of the K-axiom revealed that such tautologies are often necessary in order to get the players' reasoning off the ground. It is precisely the function of the necessitation rule to ensure that players have sufficient tautological knowledge.

2.2.2.2 Motivation of the Axioms

A pure strategy is, by definition, strictly dominated whenever there is some mixed strategy that is strictly better against all i-deleted profiles of pure strategies of his opponents. This definition allows for cases in which some pure action is only strictly dominated by a mixture of strategies, and not by a pure strategy. While it may be rather obvious that rational players avoid pure strategies when there are better ones, a defence of the present rationality principle has more difficulty showing the rationale behind not playing pure strategies that are only dominated by mixed strategies—how would a rational player in fact implement playing such a mixture?

One way to argue this uses the following lemma, which returns to fulfil similar tasks in future sections.

Lemma 2.2 (Pearce, 1984) *Let $\Gamma = (I, (A_i)_i, (u_i)_i)$ be a normal form game with any number of players. Then for all players i, an action a_i is strictly dominated by some mixed strategy over A_i against $\prod_{j \neq i} A_j$ if and only if it is not a best response among A_i to some probability measure on $\prod_{j \neq i} A_j$.*

To motivate why a rational player plays no strategy that is strictly dominated amounts, given this lemma, to motivating why such a player chooses a strategy that is a best response against some probability measure. This is a possible way. Although you will only know in rare cases exactly what a player's beliefs look like, you will always know that he or she has some beliefs rather than none, and these beliefs may take a probabilistic form. Rationality in the utility maximising sense of the word means that the player chooses a best response against these beliefs, and that is precisely a strategy that is strictly undominated. This argument assumes that players always have probabilistic beliefs. Without such beliefs, best responses make no sense; without best responses, the lemma could not be used; and without the lemma, the argument would remain inconclusive. But as the same holds for the definition of rationality as expected utility maximisation, this is not a serious problem for my approach. Here, the application of the lemma is harmless. In other contexts, this may not be the case, particularly if we use a similar lemma about weak dominance simultaneously.

2.2.2.3 Variants

The set-up of the axiom schemas is general enough to obtain a number of variants of the characterisation theorem with little extra work, and the most obvious candidate is to adopt a stronger rationality concept that not only avoids playing strictly dominated strategies, but also weakly dominated ones. A pure strategy $a_i \in A_i$ is weakly dominated by a mixed strategy σ_i against all pure opponent strategies in $\prod_{j \neq i} A_j$ if and only if

$$u_i(\sigma_i, a_{-i}) \geq u_i(a_i, a_{-i})$$

for all i-deleted action profiles $a_{-i} \in \prod_{j \neq i} A_j$, and

$$u_i(\sigma_i, a_{-i}) > u_i(a_i, a_{-i})$$

for at least one such i-deleted action profile a_{-i}. A strategy is *weakly dominated* in a game whenever there is such a mixed strategy that weakly dominates it. Notation

$$\mathrm{nwd}_i(X_i, X_{-i})$$

stands for those actions of player i that are weakly undominated in the game spanned by X_i and $\prod_{j \neq i} X_j$. Substituting in the epistemic characterisation result of iterated strict dominance any nsd_i by nwd_i, a characterisation result of iterated weak dominanceis obtained, or so it seems.

	2_1	2_2
1_1	(1,1)	(0,0)
1_2	(1,1)	(2,1)
1_3	(0,0)	(2,1)

Fig. 2.3

But that conclusion is too rash. While the outcome of iterated elimination of strictly dominated strategies is relatively independent of what in the algorithm is considered a round of elimination, slightly different algorithms for iterated weak dominance yield crucially different outcomes. In a round of elimination some, or all, weakly dominated strategies can be eliminated for some, or all, players—and the permutations of *some* and *all* give at least four different algorithms and outcomes. The standard response is that in one round all weakly dominated strategies for all players are removed.[19] This settles the definition of the game-theoretic concept, but it does not make the logical formalisation task an easy one. To see why, consider the

[19] Fudenberg and Tirole, op. cit. 461. See also Krzysztof Apt, 'The Many Faces of Rationalizability', *The B.E. Journal of Theoretical Economics*, 7 (2007), article 18.

game shown in Figure 2.3.[20] The standard procedure removes 1_1 and 1_3 in the first round, and nothing in the second, resulting in the set $\{(1_2, 2_1), (1_2, 2_2)\}$ of iteratively weakly undominated strategy profiles. The analogues of the axioms for the basis and inductive step of the characterisation of iterated strict dominance are

$$\mathbf{rat}_i \to \mathrm{nwd}_i(A_i, A_j)$$

and

$$(\mathbf{rat}_i \wedge \Box_i X_i \wedge \Box_i X_j) \to \mathrm{nwd}_i(X_i, X_j).$$

For the game shown in Figure 2.3,

$$\mathbf{rat}_1 \to \mathbf{1}_2$$

is a particular instance of such an axiom. It trivially entails

$$\mathbf{rat}_1 \to (\mathbf{1}_1 \vee \mathbf{1}_2),$$

however, which means that if it is assumed $\Box_2 \mathbf{rat}_1$, it follows that $\Box_2(\mathbf{1}_1 \vee \mathbf{1}_2)$. Since player 2's first strategy is weakly dominated in the subgame spanned by both players' first and second strategies (the entire game minus 1_3) an instance of the inductive axiom is

$$(\mathbf{rat}_2 \wedge \Box_2(\mathbf{2}_1 \vee \mathbf{2}_2) \wedge \Box_2(\mathbf{1}_1 \vee \mathbf{1}_2)) \to \mathbf{2}_1.$$

But this suggests that from the epistemic assumptions of the characterisation result (common knowledge of the fact that players play weakly undominated strategies) we can derive that player 2 does not play his first strategy, even though that strategy does survive the iterated weak dominance. Consequently, the present formalisation cannot be right. In fact, the present proposal is plainly inconsistent—witness what happens if we go through a second derivation along the lines of the first one, but with 1_3 instead of 1_1 to obtain 2_2. Combining both derivations yields a contradiction.

A way out of this predicament is to include in the consequent of the axiom not only information about the strategies an agent considers possible, but also to refer to information about which strategies an agent explicitly excludes. The basis axiom for a game-playing situation of the current game is, for instance,

$$\mathbf{rat}_1 \to (\mathbf{1}_2 \wedge \neg(\mathbf{1}_1 \vee \mathbf{1}_3)).$$

If $\overline{X_i}$ denotes the complement $A_i \backslash X_i$ of X_i (the set containing proposition letters for player i's strategies that are not in X_i) and if $\neg X$ abbreviates $\neg \bigvee X$, then axioms

$$\mathbf{rat}_i \to (\mathrm{nwd}_i(A_i, A_j) \wedge \neg\overline{\mathrm{nwd}_i(A_i, A_j)})$$

and

[20] Martin Osborne and Ariel Rubinstein, *A Course in Game Theory* (Cambridge: MIT Press, 1994), 63. A different removal procedure is used, though.

$$(\mathbf{rat}_i \wedge \Box_i(X_i \wedge \neg\overline{X_i}) \wedge \Box_i(X_j \wedge \neg\overline{X_j})) \rightarrow (\text{nwd}_i(X_i, X_j) \wedge \neg\overline{\text{nwd}_i(X_i, X_j)})$$

can be used to characterise iterated weak dominance as in the standard algorithm.

The mathematical issue of the definition of algorithms, and the logical issue of the axiomatisation rationality may be easily solvable, but this does not guarantee that the axioms involving weak dominance are plausible. It is natural to attempt to defend their plausibility on the basis of an analogue of Lemma 2.2.

Lemma 2.3 (Pearce, 1984) *Let $\Gamma = (I, (A_i)_i, (u_i)_i)$ be an N-person normal form game. Then for all i, an action a_i is weakly dominated by some mixed strategy over A_i against $\prod_{j \neq i} A_j$ if and only if it is not a best response among A_i to some probability measure with full support over $\prod_{j \neq i} A_j$.*

But as this lemma requires full-support beliefs assigning zero probability to no alternative, this approach seems blocked. To assume, as in the motivation underlying strict dominance, that players possess just one probabilistic belief or another seems quite reasonable. To presume, as has to be done here, that their beliefs are such that they do not assign zero probability to anything is not generally plausible. An important goal of forming beliefs is to exclude certain things from consideration by assigning them zero probability. Other approaches, investigated later in the context of mixed iterated strict weak dominance and the Dekel–Fudenberg procedure, use lexicographic beliefs, but while this helps to support the basis axiom, iterating this to subgames is quite problematic. I will therefore put the plausibility of weak dominance to one side until the discussion of mixed iterated strict weak dominance and the Dekel–Fudenberg procedure in Section 2.4.[21]

Before proceeding to another way of generalising the epistemic characterisation theorem of iterated strict dominance, I will examine why the logical problem does not arise (there is no issue here, the different algorithms just give the same outcome). In short, the answer is that nsd_i are monotone, but the nwd_i not.

Remark 2.1 If $X_i \subseteq A_i$ and $X_j \subseteq Y_j \subseteq A_j$, then $\text{nsd}_i(X_i, X_j) \subseteq \text{nsd}_i(X_i, Y_j)$.

Remark 2.2 If $Y_i \subseteq X_i \subseteq A_i$ and $X_j \subseteq A_j$, then $\text{nsd}_i(X_i, X_j) \cap Y_i \subseteq \text{nsd}_i(Y_i, X_j)$.

Suppose, for example, that player 1 believes that 2 will play (from four possible strategies) the first or second one. This is most naturally formalised as $\Box_1(\mathbf{2}_1 \vee \mathbf{2}_2)$. In an attempt to copy the problematic disjunction from above, player 1 might consider the larger set, and phrase her beliefs as $\Box_1(\mathbf{2}_1 \vee \mathbf{2}_2 \vee \mathbf{2}_3 \vee \mathbf{2}_4)$. But while a contradiction followed in the case of iterated weak dominance, in the proof system for rationality in terms of strict dominance this is not the case. Player 1's belief is less informative, but since the rationality concept at stake is monotone, what follows

[21] Adam Brandenburger, Amanda Friedenberg and H. Jerome Keisler, 'Admissibility in Games', *Econometrica*, 76 (2008), 307–352 provide a sophisticated approach to iterated weak dominance, taking lexicographic probability systems as modelling instruments. Larry Samuelson, 'Dominated Strategies and Common Knowledge', *Games and Economic Behavior*, 4 (1992), 284–313 suggests that common knowledge of the fact that players do not play weakly dominated strategies does not guarantee that the outcome is iteratively weakly undominated because such common knowledge entails inconsistencies in the beliefs of the players.

from the less informative beliefs is consistent with what would follow from the more refined beliefs.

2.2.2.4 More than Two Players

The result is proven for two-person normal form games, but it can easily be extended to arbitrary numbers of players. Player i's beliefs about her opponents are in that case phrased as $\Box_i(\bigvee X_1 \wedge \cdots \wedge \bigvee X_{i-1} \wedge \bigvee X_{i+1} \wedge \cdots \wedge \bigvee X_n)$ instead of as $\Box_i \bigvee X_j$, and the nsd$_i$ will be defined on pairs of the form $(X_i, (X_j)_{j \neq i})$ where $X_i \subseteq A_i$ and $X_j \subseteq A_j$. But while this is a mathematically and logically trivial move, problems arise at the level of plausibility. As I have demonstrated, the two-person case inherits its plausibility from Lemma 2.2. Assuming that any player possesses probabilistic beliefs, it is clear that if she is rational, she chooses a best response given those beliefs, and such a best response, the lemma shows, is strictly undominated.

The argument hinges on the assumption that the players have probabilistic beliefs. If you assume that someone has probabilistic beliefs, then the lemma helps to show that she avoids playing strictly dominated strategies, but if you do not wish to make that assumption, no such line of reasoning is possible. The probabilistic beliefs required by the lemma are, however, elements of $\Delta \prod_{j \neq i} X_j$. In a two-person game-playing situation, players are supposed to have probability measures on the opponent's actions. But in situations involving more than two players, they are supposed to have probability measures on i-deleted strategy profiles, rather than a collection of $N-1$ probability distributions on each individual opponent. One instead of $N-1$ means that players possibly assign correlation to their opponents, and this contradicts the ban on exogenous information. If a player believes that the actions of two of her opponents are correlated (say they both play their first strategies with probability one-half, they play their second strategies with probability one-half, thus giving zero probability to the event that the one plays his first and the other his second strategy) then she cannot justify these beliefs solely on the basis of the game matrix. An appeal to exogenous information, that is, cannot be avoided if one wishes to allow for correlated beliefs.[22]

For a concrete example consider the three-person normal form game shown in Figure 2.4.[23] The first player chooses rows (1_1 for the upper row, 1_2 for the lower), the second columns (2_1 for the left column, 2_2 for the right), the third matrices (enumerated starting from the left), and they all earn the same payoff. The second matrix from the left is not strictly dominated. If $\mathsf{P}_3((1_1,2_1)) = \mathsf{P}_3((1_2,2_2)) = \frac{1}{2}$, then the expected utilities satisfy $\mathrm{EU}_3(\mathsf{P}_3,3_1) = \mathrm{EU}_3(\mathsf{P}_3,3_2) = \mathrm{EU}_3(\mathsf{P}_3,3_3) = 4 >$

[22] Robert Aumann, 'Correlated Equilibrium as an Expression of Bayesian Rationality', *Econometrica*, 55 (1987), 16 gives a standard example where the opponents are graduates from the same business school, possibly responding in similar ways to situations that were studied in school. That two of your opponents went to the same school, however, is not part of the game structure here, and therefore the correlation is derived exogenously. For further discussion of the correlated equilibrium, see Section 4.2.2.2.

[23] Osborne and Rubinstein, op. cit. 57–58.

8	0
0	0

4	0
0	4

0	0
0	8

3	3
3	3

Fig. 2.4

$3 = \mathrm{EU}_3(\mathsf{P}_3, 3_4)$. Consequently, the second matrix is a best response to P_3. But if P_3 counts as a belief of player 3 about his opponents, he believes that his opponents coordinate in some way, and it is impossible to attribute alternative beliefs to player 2 that would not ascribe correlation to 1 and 2. If player 3 believes that 1 plays 1_1 with probability p (and 1_2 with $1-p$), and that 2 plays 2_1 with probability q (and 2_1 with $1-q$) then the second matrix from the left would be a best response if and only if

$$4pq + 4(1-p)(1-q) \leq 8pq, 8(1-p)(1-q), 3.$$

This is a system of inequalities without a solution, however. Correlated beliefs may be a rather common phenomenon in many cases of strategic interaction, but they are excluded by the ban on exogenous information, and consequently the solution concept of iterated strict dominance loses its attraction for more than two players.[24]

2.2.2.5 Stalnaker's Game Models Approach

While most game theorists have applied probability theory to detect the epistemic foundations of iterated strict dominance, Robert Stalnaker has proposed an original and fruitful way to apply the model theory of epistemic logic.[25] To bring to light

[24] The issue is subtle, though. Adam Brandenburger and Amanda Friedenberg, 'Intrinsic Correlation in Games', *Journal of Economic Theory*, 141 (2008), 28–67 propose to derive correlation endogenously from players' beliefs that their common beliefs are correlated, but seem to be in need of some assumption of correlation on hierarchies of beliefs.

[25] 'On the Evaluation of Solution Concepts', *Theory and Decision*, 37 (1994), 49–73 deals with iterated strict dominance. See also 'Knowledge, Belief and Counterfactual Reasoning in Games', *Economics and Philosophy*, 12 (1996), 133–163 (repr. with proofs in C. Bicchieri, R. Jeffrey and B. Skyrms (eds.), *The Logic of Strategy* (New York: Oxford University Press, 1999), 3–38), 'Belief

the tacit assumptions about players' reasoning processes underlying various game-theoretic models, the approach I have chosen is syntactic rather than semantic; it remains close in spirit to many of Stalnaker's ideas, though. The remainder of this section compares the two.

One of Stalnaker's innovations is to interpret epistemic characterisation results as relations between classes of game models and solution concepts. A *game model* for an N-person normal form game $\Gamma = (I, (A_i)_i, (u_i)_i)$ is a model $M_\Gamma = (W, (R_i)_i, (\mathsf{P}_i)_i, (s_i)_i)$ for an epistemic language (with probabilistic beliefs) together with the indication of the actual world **a** of that model. The R_i are serial, transitive and Euclidean accessibility relations (corresponding to the D-axiom, the 4-axiom and the 5-axiom, respectively), the $\mathsf{P}_i(w)$ capture the probability measure of player i at world w, and $s_i(w)$ the strategy i performs at world w. Common belief is defined as usual, and as the accessibility relations are not required to be reflexive, beliefs may be false. A player i is said to be rational at some world v if and only if for all $a_i \in A_i$ it holds that

$$\sum_{w \in W} \mathsf{P}_i(v)(\{w\}) \cdot u_i(s_1(w), \ldots, s_i(w), \ldots, s_n(w)) \geq$$
$$\sum_{w \in W} \mathsf{P}_i(v)(\{w\}) \cdot u_i(s_1(w), \ldots, a_i, \ldots, s_n(w)).$$

Given a game Γ let $S \subseteq \prod_i A_i$ be the set of strategy profiles singled out by a solution concept. A class M of game models for Γ characterises the solution concept whenever, first, for any model in M there is a strategy profile in S that is played in the actual world of the model, and, second, for every strategy profile in S there is a model in M in the actual world of which the strategy profile is played. Stalnaker's version of the epistemic characterisation result for iterated strict dominance is the following theorem.

Theorem 2.4 (Stalnaker, 1994) *Let $\Gamma = (I, (A_i)_i, (u_i)_i)$ be an N-person normal form game. Then iterated strict dominance is characterised by the class of game models in which there is common true belief among the players about the utility functions of all players and it is common true belief among them that they are rational.*

The proof Stalnaker offers makes essential use of an equivalent formulation of iterated strict dominance using two lemmas about the solution concept of correlated rationalisability. Before stating the lemmas, let me first clarify this notion. Set $\mathsf{R}_i^0 = \Delta(A_i)$, the set of player i's mixed strategies. Then define recursively sets R_i^{n+1} to contain those elements (mixed strategies) σ_i from R_i^n for which there is some i-deleted profile of mixed strategies $\sigma_{-i} \in \prod_{j \neq i} \mathrm{hull}(\mathsf{R}_j^n)$—the convex hull , that is, the smallest convex superset—such that

$$u_i(\sigma_i, \sigma_{-i}) \geq u_i(\sigma_i', \sigma_{-i})$$

Revision in Games: Forward and Backward Induction', *Mathematical Social Sciences*, 36 (1998), 31–56, and 'Extensive and Strategic Forms: Games and Models for Games', *Research in Economics*, 53 (1999), 293–319.

for all $\sigma_i' \in \mathsf{R}_i^n$. The set $\mathsf{R}_i^\infty = \bigcap_n \mathsf{R}_i^n$ is then the set of (non-correlated) *rationalisable* strategies of player i. The idea behind this solution concept is that players choose a (possibly mixed) action that is a best response to profiles of mixed actions of their opponents. Players all see that their opponents do that, and hence they can remove strategies that fail to be best responses against profiles of opponents' strategies in, first, the entire game, and, subsequently, in successively smaller subgames. The players' beliefs do not correlate actions of opponents here, but beliefs allowing for correlation are easily represented if only one probability measure on the product of i-deleted mixed rationalisable strategy profiles is used in the definition, instead of $N-1$. Modifying the above definition of rationalisability to allow for such beliefs yields *correlated rationalisability*.

Lemma 2.4 *Let $\Gamma = (I, (A_i)_i, (u_i)_i)$ be an N-person normal form game. A pure action $a \in A_i$ is correlated rationalisable if and only if for each player j there is a set $X_j \subseteq A_j$ such that, first, $a \in X_i$, and second, for all players j and all actions $b \in X_j$ there is a probability measure $\mathsf{P}_{i,b}$ over $\prod_{k \neq j} X_k$ to which b is a best response among X_j.*

Lemma 2.5 (Pearce, 1984) *Let $\Gamma = (I, (A_i)_i, (u_i)_i)$ be an N-person normal form game. Then the set of correlated rationalisable action profiles and the set of the iteratively undominated profiles are identical.*

I can now prove Stalnaker's version of the epistemic characterisation theorem of iterated strict dominance.

Proof One direction is proven. Set

$$X_i = \{a \in A_i \mid s_i(w) = a, \text{ and } R_{\mathbf{C}}(\mathbf{a}, w) \text{ or } w = \mathbf{a}\},$$

where $R_{\mathbf{C}}$ is the accessibility relation for the **C**-operator, that is, the transitive, but not necessarily reflexive, closure of the union of the R_i. This is the set of those strategies that are not commonly excluded, so to speak. It is readily recognisable that the X_i satisfy the conditions of Lemma 2.4. Let a be some element of X_i. Because of common belief about rationality (and, for $w = \mathbf{a}$, because of rationality) a is played by i in a world w, say, where i is rational. That means, however, that relative to the probabilistic beliefs $\mathsf{P}_i(w)$ the action a is a best response (among X_i). Owing to the set-up of the sets X_j, $j \neq i$, the support of $\mathsf{P}_i(w)$ is a subset of $\prod_{j \neq i} X_j$. Applying Lemmas 2.4 and 2.5, and observing that the strategy i plays in $\mathbf{a}$ is a member of X_i, concludes the proof.

What this overview immediately lays bare is that the game model approach makes it readily possible to prove an additional direction of the epistemic characterisation result. It is shown not only that under certain epistemic assumptions the solution concept of iterative strict dominance obtains, but also that for any iteratively undominated strategy profile there is a game-playing situation in which these precise epistemic assumptions hold true. This is provable in my system, too, but it is much more transparent in Stalnaker's.

A more theoretically relevant difference between the approaches concerns the way in which rationality is formulated. Where Stalnaker adopts the explicit definition of rationality as expected utility maximisation, I propose an inductive and implicit one that, I believe, is preferable given my interpretative goals. Drawing a line between a basis case without beliefs and an inductive case with beliefs makes it possible to mimic every round of elimination of the solution concept by a step in the hierarchy of common beliefs about rationality, and this is reflected by the decidedly inductive character of the proof—in contrast to Stalnaker's. One could say that the reference to Lemma 2.2, while making it possible to discuss directly the plausibility of the notions used, suppresses in Stalnaker's approach the procedural issues of the solution concept and the way subsequent steps in the removal procedure are related to subsequent levels of common belief. This also becomes clear when experimental findings on game-playing are considered. Many experiments suggest that actual players of normal form games do not go through more than three rounds of elimination of dominated strategies, and my axiomatisation of rationality allows me to relate their reasoning processes (epistemic assumptions) to their decisions (solution concept) in a highly explicit fashion.[26] In addition, the bare logical form of my formalism easily suggests various extensions of the epistemic characterisation result that, using game models, would come to mind less readily. It is now time to investigate another variant, the Dekel–Fudenberg procedure.

2.3 The Dekel–Fudenberg Procedure

2.3.1 The Epistemic Characterisation Theorem

Common true belief about utility structure and perfect rationality entails the Dekel–Fudenberg procedure of one round of elimination of weakly dominated strategies followed by infinitely many rounds of elimination of strictly dominated strategies.[27]

The solution concept is, with the usual slight abuse of notation, captured by

---

[26] Colin Camerer, *Behavioral Game Theory: Experiments in Strategic Interaction* (Princeton: Princeton University Press, 2003), 199–264.

[27] Robert Stalnaker, 'Knowledge, Belief and Counterfactual Reasoning in Games', *Economics and Philosophy*, 12 (1996), 133–163 (repr. with proofs in C. Bicchieri, R. Jeffrey and B. Skyrms (eds.), *The Logic of Strategy* (New York: Oxford UP, 1999), 3–38), and 'Belief Revision in Games: Forward and Backward Induction', *Mathematical Social Sciences*, 36 (1998), 31–56. Stalnaker also provides a characterisation of a variant with two rounds of elimination of weakly dominated strategies. The Dekel–Fudenberg procedure was developed by Eddie Dekel and Drew Fudenberg, 'Rational Behavior with Payoff Uncertainty', *Journal of Economic Theory*, 70 (1990), 243–267, who characterised the concept in terms of payoff-uncertainty rather than perfect rationality. See Section 2.4.

$$\begin{aligned} \mathsf{DF}_i^0 &= A_i, \\ \mathsf{DF}_i^1 &= \mathrm{nwd}_i(\mathsf{DF}_1^0,\ldots,\mathsf{DF}_N^0), \\ \mathsf{DF}_i^{n+1} &= \mathrm{nsd}_i(\mathsf{DF}_1^n,\ldots,\mathsf{DF}_N^n) \quad (n \geq 1). \end{aligned}$$

The concept of *perfect rationality* is more involved. Robert Stalnaker developed it to mirror the fact that in order to decide on an action, players may use not only their actual beliefs, but also the beliefs they would adopt if they were to learn that their actual beliefs are false. If, given my actual beliefs, more than one action maximises expected utility, I can winnow down my choice by inspecting which of the actions would still be optimal if I changed my beliefs. As a perfectly rational player, I would choose an action that remains after the sifting.

	2_1	2_2	2_3
1_1	(1,4)	(1,8)	(1,0)
1_2	(5,8)	(1,4)	(3,0)
1_3	(3,4)	(0,8)	(2,0)

Fig. 2.5

In order to see this in a context of game-playing, consider the game shown in Figure 2.5. Clearly, the third strategy of player 1 is a bad choice—no matter what she believes about her opponent. Now suppose that for whatever reason she has adopted the belief that player 2 plays 2_2. Then two utility maximising strategies remain, namely, 1_1 and 1_2. So as to decide between them, player 1 checks how she would change her beliefs if she learned that her belief about 2_2 is wrong, finding, for the sake of argument, that she would then believe that her opponent plays 2_1. While given her belief about 2_2 it makes no difference whether she plays 1_1 or 1_2, given her (hypothetical) belief about 2_1, only 1_2 maximises expected utility. If player 1 were merely rational, she could play 1_1 or 1_2. But as she is perfectly rational, she only plays 1_2.

2.3.1.1 An Implicit, Inductive Formalisation of Perfect Rationality

The way in which I formalise the notion of perfect rationality differs significantly from Stalnaker's formalisation. Using an implicit, inductive axiomatisation characteristic of the syntactic approach to the Epistemic Programme proffered here, the basic axiom requires perfectly rational players to avoid playing *weakly* dominated strategies,

PRat$_{bas}$ $\quad \mathbf{prat}_i \rightarrow \mathrm{nwd}_i(A_i, A_j)$,

while the inductive case excludes strategies that are *strictly* dominated in the subgame the players believe to be played,

PRat$_{ind}$ $\quad (\mathbf{prat}_i \wedge \Box_i X_i \wedge \Box_i X_j) \rightarrow \mathrm{nsd}_i(X_i, X_j)$.

Table 2.5

Assumptions	
preferences	$\bigwedge_{i,k,l} \mathbf{u}_i(k,l) = \mathbf{r}_{i,k,l}$
principles	$\bigwedge_i \mathbf{prat}_i$
beliefs	
preferences	$\mathbf{C} \bigwedge_{i,k,l} \mathbf{u}_i(k,l) = \mathbf{r}_{i,k,l}$
principles	$\mathbf{C} \bigwedge_i \mathbf{prat}_i$
performed action	–
Solution Concept	
player i	DF_i^∞
Proof System	$_\Gamma\mathbf{K_{EC}prat}$

A proof system $_\Gamma\mathbf{K_{EC}prat}$ is used with axioms Prop, Dual, K, E, C, the proof rules modus ponens, necessitation and induction, the first four axioms for normal form game-playing situations, plus the two axioms for perfect rationality. The assumptions of the characterisation theorem are formalised in Table 2.5, where $\mathbf{prat}_i$ stands for perfect rationality, DF_i^n for player i's strategies that survive n rounds of the Dekel–Fudenberg procedure, and DF_i^∞ for the limit.

Theorem 2.5 (Stalnaker, 1996) *Let Γ be a two-person normal form game. Assume that the following two conditions are satisfied.*

1. *There is common knowledge among the players about the utility functions of all players.*
2. *It is common knowledge among the players that they are perfectly rational.*

Then the players play strategies that survive the Dekel–Fudenberg procedure, that is, the iterated elimination of strictly dominated strategies after one round of elimination of weakly dominated strategies.

Proof It is demonstrated that

$$\forall n \vdash (\mathbf{prat}_i \wedge \Box_i \mathbf{E}^{\leq n-1}(\mathbf{prat}_i \wedge \mathbf{prat}_j)) \rightarrow DF_i^{n+1}, \tag{2.1}$$

Define φ_i^0 as $\mathbf{prat}_i$, and φ_i^{n+1} as $\varphi_i^n \wedge \Box_i(\varphi_i^n \wedge \varphi_j^n)$. It is easy to see that φ_i^n is provably equivalent to the antecedent of 2.1. Now I can turn to the proof of

$$\forall n \vdash \varphi_i^n \rightarrow DF_i^{n+1}.$$

The case of $n = 0$ is an instance of PRat_{bas}. To prove the inductive step assume as inductive hypothesis

$$\varphi_i^n \to DF_i^{n+1},$$

and similarly for j. Applying necessitation for i to both inductive hypotheses yields

$$\Box_i(\varphi_i^n \wedge \varphi_j^n) \to (\Box_i DF_i^{n+1} \wedge \Box_i DF_j^{n+1}),$$

from which, by definition, it follows that

$$\varphi_i^{n+1} \to (\mathbf{prat}_i \wedge \Box_i DF_i^{n+1} \wedge \Box_i DF_j^{n+1}).$$

Applying the PRat_{ind}-axiom yields

$$\varphi_i^{n+1} \to \mathrm{nsd}_i(DF_i^{n+1}, DF_j^{n+1}),$$

which concludes the proof.

2.3.2 *Discussion*

2.3.2.1 Stalnaker's Game Models Approach

Robert Stalnaker approaches this theorem using the game models I discussed in the previous section. To capture perfect rationality, the framework has to be enlarged to make it possible to reason about belief revision. I present the idea here by means of a short excursion to conditional logic. This contrasts slightly with Stalnaker's own presentation, but ultimately yields the same result.

The first step in the presentation is to add to the logical language symbols $>_i$ with well-formed formulae $\varphi >_i \psi$ with interpretation 'if player i learned φ, she would believe ψ'.[28] A standard model for conditional logic in a multi-agent setting is a triple $M = (W, (f_i)_i, v)$ where W is the set of possible worlds, $f_i \colon W \times \wp(W) \to \wp(W)$ a *selection* function and v a *valuation* function as usual. Truth for sentences involving the $>_i$-sign is defined as

$$M, w \models \varphi >_i \psi \text{ if and only if } f_i(w, [\varphi]) \subseteq [\psi].$$

The underlying idea is that game models for perfect rationality represent belief revision in the following way. While in every world w of a game model, player i's usual beliefs can be obtained by simply inspecting the R_i-arrows that start from w, giving extra structure to the game model allows us, in addition, to read off from the game model the players' dispositions to belief revision at w, and sentences $\varphi >_i \psi$, using conditional logic, are introduced precisely to that end. They express belief revision

[28] In conditional logic, the interpretation is not in terms of belief revision, but in terms of counterfactual implications.

dispositions in the sense that, if true at w, player i would revise her beliefs to ψ if she learned that φ. That being the case, the main challenge is to define the selection function to give $>_i$ precisely this desired interpretation, and that is what I will turn to first.

To start with, belief revision ought to satisfy certain conditions of theoretic rationality, and an obvious answer to this requirement is the set of standard AGM-postulates for belief revision.[29] These postulates are conditions on revision functions * that map sets of worlds plus (extensions of) sentences to sets of worlds with the interpretation that if B is the initial set of worlds a player believes possible then $^*(B,[\varphi])$ is the set of worlds she holds possible after she has learned that φ is the case. Notation $[\varphi]$ is used for the extension of φ, and B^*_X for $^*(B,X)$, and applied to game models the AGM-conditions are the following.[30]

$$B^*_X \subseteq X.$$
$$\text{If } X \neq \emptyset, \text{ then } B^*_X \neq \emptyset.$$
$$\text{If } B \cap X \neq \emptyset, \text{ then } B^*_X = B \cap X.$$
$$\text{If } B^*_X \cap Y \neq \emptyset, \text{ then } B^*_{X \cap Y} = B^*_X \cap Y.$$

The first of these postulates states that if a player revises her current beliefs B with new information X, she ends up with an epistemic state in which X is true. The second requires that new information will never lead a player to draw inconsistent conclusions unless the information itself is inconsistent. The third states the rather obvious practice that if the current beliefs B and the new information X have a non-empty intersection, the new beliefs are formed by the intersection. The last AGM-postulate, finally, is more difficult. It states that there must be no difference between subsequently revising beliefs B with X and Y, and simultaneously revising beliefs B with $X \cap Y$—unless the intermediate step B^*_X of the subsequent revision process has nothing in common with Y.

Using the functions satisfying the AGM-postulates the desired selection function can now be defined as

$$f_i(w,[\varphi]) = (B^*_{i,w})_{[\varphi]}$$

where $B_{i,w} = \{v \in W \mid R_i(w,v)\}$ represents player i's beliefs at w. It is, however, no trivial matter to show that this is well-defined, and propositions are needed to demonstrate the existence of the * and the existence of the right B.

Proposition 2.1 *Let Q be a reflexive, transitive and connected relation on some set W. Then there is a revision function $B^*: \wp(W) \to \wp(W)$ for some $B \subseteq W$ satisfying the AGM-conditions for all $X \subseteq W$.*

Proof It is left to the reader to verify that putting

[29] Carlos Alchourrón, Peter Gärdenfors and David Makinson, 'On the Logic of Theory Change: Partial Meet Functions for Contraction and Revision', *Journal of Symbolic Logic*, 50 (1985), 510–530.

[30] The conditions are stated not for revision with sentences but with propositions or sets of possible worlds X and Y, but this is immaterial.

$$B = \{w \in W \mid Q(v,w) \text{ for all } v \in W\}$$

and

$$B^*_X = \{w \in X \mid Q(v,w) \text{ for all } v \in X\}$$

works.

This proposition suggests that once Stalnaker's original game models are decorated with such Q_i-relations for every player i there is a representation of dispositions to revise beliefs for some specific belief set $B \subseteq W$. So as to obtain a representation for $B_{i,w}$, another proposition is needed. Here the *belief set* $B_{i,w}$ of player i at w is the set of worlds accessible from w for i, and if the belief sets of two worlds v and w coincide (that is, if at those worlds i's beliefs are identical) the worlds are *subjectively indistinguishable* for i, with notation $T_{i,w}$ for the worlds subjectively indistinguishable from w for i. Recall that the accessibility relations for the beliefs are transitive and Euclidean (they are also serial, but that is not needed here). It follows from $R_i(w,v)$ that w and v are subjectively indistinguishable, which establishes the following inclusion relation used in the proof of the first proposition.

Remark 2.3 For all i and w it is true that $B_{i,w} \subseteq T_{i,w}$.

Proposition 2.2 *Let R_i be serial, transitive and Euclidean on W. Let Q_i be reflexive, transitive, and connected on each set $T_{i,w}$ such that if two worlds are connected they are subjectively indistinguishable. Assume that $R_i(w,v)$ if and only if*

$$\forall w(Q_i(w,u) \vee Q_i(u,w)) \rightarrow Q_i(u,v)).$$

Then it is true that

$$B_{i,w} = \{u \in T_{i,w} \mid Q_i(v,u) \textit{ for all } v \in T_{i,w}\}.$$

Proof The one direction follows from transitivity and Euclideanness of R_i. The other direction is an immediate consequence of the assumption.

This gives the desired way to decorate the original game models with the Q_i-relations. If the condition relating the R_i and the Q_i is satisfied, then we can interpret sentences from the earlier conditional language on a model of the form tuple $(W,(R_i)_i,(Q_i)_i,(\mathsf{P}_i)_i,(s_i)_i)$. Note that what the interrelation condition really says is that the elements v that stand in the R_i-relation to w are exactly the elements that stand in the R_i-relation to any element in $T_{i,w}$.

This detour via conditional logic makes it possible to define perfect rationality. To define that notion is to define what it means to make mistakes.[31] Set recursively

$$\mathrm{ERR}^1_i = \{w \in W \mid \text{ not } R_i(w,w)\},$$

the set of worlds in which i is in error, and

[31] The definition deviates slightly from Stalnaker, 'Knowledge, Belief and Counterfactual Reasoning in Games', 147, but is equivalent.

$$\mathrm{ERR}_i^{n+1} = \{w \in \mathrm{ERR}_i^n \mid w \notin B_{i,w}^*(\mathrm{ERR}_i^n)\},$$

the set of worlds that do not survive revision with the proposition ERR_i^n. So ERR_i^2, for instance, is the set where i is in error, and even if i would revise her beliefs with the proposition that she is in error, she would still be in error. Omitting the discussion about conditionalisation on probability zero events, set

$$\mathsf{P}_i([\varphi] \mid [\psi]) = \frac{\mathsf{P}_i([\varphi] \cap B_{i,w}^*([\varphi]))}{\mathsf{P}_i(B_{i,w}^*([\psi]))},$$

which for $\mathsf{P}_i([\psi]) > 0$ is the same as Bayesian updating. Then, set inductively

$$\mathrm{PRAT}_{i,w}^0 = \{a_i \in A_i \mid \mathrm{EU}_{i,w}(a_i) \geq \mathrm{EU}_{i,w}(b_i) \text{ for all } b_i \in A_i\},$$

the set of (not necessarily perfectly) rational actions, and $\mathrm{PRAT}_{i,w}^{n+1} =$

$$\{a_i \in \mathrm{PRAT}_{i,w}^n \mid \mathrm{EU}_{i,w}(a_i \mid \mathrm{ERR}_i^{n+1}) \geq \mathrm{EU}_{i,w}(b_i \mid \mathrm{ERR}_i^{n+1}) \text{ for all } b_i \in \mathrm{PRAT}_{i,w}^n\},$$

where the conditional expected utility is defined by

$$\mathrm{EU}_{i,w}(a_i \mid [\varphi]) = \sum_{a_{-i} \in A_{-i}} \mathsf{P}_i(S^{-1}[a_{-i}] \mid [\varphi]) \cdot u_i(a_i, a_{-i}),$$

and the non-conditional expected utility $\mathrm{EU}_{i,w}(a_i)$ in an obvious analogous way. Finally, the set $\mathrm{PRAT}_i^\infty = \bigcap_n \mathrm{PRAT}_i^n$ is the set of perfectly rational strategies for player i at w.

If this is Stalnaker's conception of perfect rationality, how do the proposition letters $\mathbf{prat}_i$ compare to $\mathrm{PRAT}_{i,w}$? Since my logic is not really a logic for game models in Stalnaker's sense, it does not make sense to state a genuine equivalence result relative to some class of models. Since there is no possibility of a strict comparison, I will examine Stalnaker's proof and show that what his proof needs in order to work is captured adequately in my $\mathbf{prat}_i$. Perhaps surprisingly this does yield some insight into the similarities and differences in our approaches.

Stalnaker's proof of the epistemic characterisation of the Dekel–Fudenberg procedure is similar to the proof for iterated strict dominance that I discussed in the previous section.

Proof Set $X_i = \{a \in A_i \mid s_i(w) = a, \text{ and } R_{\mathbf{C}}(\mathbf{a}, w) \text{ or } w = \mathbf{a}\}$. This is the set of strategies that lie on the common belief paths departing from the actual world, plus the strategy played in the actual world. It has to be shown that, if it is also assumed that if all such worlds are worlds where the players are perfectly rational (a member of $\bigcap_i \mathrm{PRAT}_i^\infty$) then in no such world will a strategy be chosen that does not survive the Dekel–Fudenberg procedure.

First it is shown that any member of X_i is not weakly dominated in the original game Γ. Clearly, such a strategy cannot be strictly dominated, because then i would not be rational at the respective world, but the extra demands of perfect rationality ensure that it cannot be *weakly* dominated either.

Second, consider the subgame Γ' of the original game Γ spanned by the sets $\mathrm{nwd}_i(A_i, A_j)$ for all i. It is shown that the X_i fulfil the conditions in Lemma 2.4 (correlated rationalisability) to conclude, using Lemma 2.5 (correlated rationalisability and iterated strict dominance coincide), that any member of X_i is iteratively strictly undominated in Γ'. Now let $a \in X_i$. Since a is being played in a world (say w) in which i is perfectly rational, a is certainly a best response among X_i to $\mathsf{P}_i(w)$, the support of $\mathsf{P}_i(w)$ is a subset of X_j by definition, for indeed, at w the strategies that player i considers possible for himself are the strategies played in worlds v with $R_i(w,v)$, and hence members of X_i.

It is quite clear that such reasoning can be carried out in my formalism equally well. In my formalism, too, the claim can be defended that the strategy chosen is not weakly dominated in the original game—the content of the PRat_{bas} axiom. And it can be shown equally well that in the subgame spanned as indicated in the proof, the strictly dominated strategies are shunned. This is precisely what PRat_{ind} states. The two steps taken in Stalnaker's proof are strictly mirrored by the two axioms that inductively axiomatise perfect rationality. To overstate slightly, the sentences from Stalnaker's proof have translations in my framework that constitute a proof of the statement that $\mathbf{prat} \wedge \mathbf{Cprat}$ at the actual world implies that the strategies played there survive, first, one round of elimination of weakly dominated strategies, and then just as many rounds of elimination of strictly dominated ones as necessary—provided we consider models for $_{\Gamma}\mathbf{K_{EC}prat}$. It is unimportant that these models are quite different from Stalnaker's game models.

2.3.2.2 Motivation of the Axioms

The result of all this is to have shown the similarities and differences between Stalnaker's approach and my own. Nothing is said, however, about the PRat-axioms. I will conclude with an argument about their plausibility, paying attention first and foremost to the fact that these axioms do not refer to the same rationality principle.[32] The general strategy for defence is to imagine that some player i is at some possible world w of, roughly, a game model in Stalnaker's sense. Player i's beliefs about what she and her opponent j play are represented by sets of strategies X_i and X_j respectively, and the question that needs to be answered concerns what she will play.[33]

---

[32] The presentation owes much to e-mail discussions with Adam Brandenburger and Robert Stalnaker for which I am very grateful.

[33] This is not entirely precise. If you think of X_i and X_j as minimal such sets of strategies played in the worlds v for which $R_i(w,v)$, then X_i is a singleton, because i knows what she plays. But if you think of these sets as non-minimal, this is not so, and depending on whether or not the sets are minimal, lemmas requiring full-support beliefs cannot be appealed to; non-minimal sets do not represent full-support beliefs. In order to characterise the Dekel–Fudenberg procedure, the axioms will be invoked for non-minimal sets. This is understandable from the procedural point of view, advocated here, concerning epistemic characterisations, where subsequent applications of the inductive axiom mirror the processes in the minds of the players leading them to eliminate

A natural way to start is to try and argue for the plausibility of the PRat-axioms using Lemma 2.3 (weakly undominated strategies are best responses to some necessarily full-support probabilistic belief) in the way I discussed in the previous section, for if it is established that any perfectly rational player has such beliefs, then the PRat$_{bas}$-axiom seems justified. But this way is blocked, because even if full support could be motivated, this motivation would be hard to square with using strict dominance in the PRat$_{ind}$-axiom, using beliefs that do not have full support in a similar motivation via Lemma 2.2.

Lexicographic probability systems provide a way out. The representation theorem (see the Appendix A) reveals that if one strategy weakly dominates another, the expected utility of the one is larger in the lexicographic sense than that of the other. If it can be established that any perfectly rational player has such a lexicographic probability system, then the PRat$_{bas}$-axiom seems justified—or not, for if I start interpreting some player's beliefs as lexicographic probability systems, then I ought to make this interpretation explicit in the treatment of beliefs in the PRat$_{ind}$-axiom as well, and so the problem from the first attempt returns.

This is intended to outline some of the intricacies of defending the plausibility of rationality axioms by reference to (lexicographic) beliefs, but it is also a good starting point to develop a fully functional argument in favour of the axioms. Let me start with the PRat$_{ind}$-axiom. Consider a perfectly rational player i in some possible world w. Because of her rationality, her choice at w maximises expected utility given her beliefs $\Box_i X_i$ and $\Box_i X_j$. Using Lemma 2.2 (strictly undominated strategies are best responses to some not necessarily full-support probabilistic beliefs) it transpires that she does not choose a strategy that is strictly dominated in the subgame spanned by X_i and X_j, and that is exactly the content of PRat$_{ind}$.

Turning to the PRat$_{bas}$-axiom, that player i is perfectly rational also entails that at w she possesses a hierarchy of dispositions to revise her beliefs that she uses in a tie-breaking procedure; what she chooses at w maximises expected utility given an initial segment of the hierarchy until no tie shows up, precisely as in the example with which this section started. The representation theorem (see the Appendix A) entails that this strategy, in addition to being strictly undominated in the subgame spanned by X_i and X_j, is *weakly* undominated in the *original* game, for it is to the effect that if she is perfectly rational with such lexicographic beliefs, the agent never plays a strategy that is not lexicographically optimal. There is a caveat, though, for the optimal strategy may be a mixed action, and therefore an extra assumption about the epistemic subject is necessary. It is assumed that there is no alternative that i assigns zero probability to in all stages of the hierarchy of her lexicographic beliefs. In other words, for any alternative there is some stage of the hierarchy at which i holds it positively epistemically open.[34]

dominated strategies. The sets get smaller and smaller during this reasoning process, but not until they have reached the final conclusion will the sets be minimal. For that reason, the argument for the plausibility of the axioms has to proceed with possibly non-minimal X_i and X_j.

[34] Stalnaker, art. cit. takes a different yet equivalent way, using closure conditions on game models that I have not dealt with in the above exposition.

While this motivates the PRat_{bas}-axiom, it remains to be seen why the nwd_i is not used in the PRat_{ind}-axiom. For an appropriate application of Lemma 2.3 to defending weak undominance, player i would need a full-support belief over the X_j. Because X_j is not necessarily minimal, from $\Box_i X_j$ no full-support belief over X_j ensues, but only a full-support belief over some subset of X_j, and on that account this lemma cannot be applied. Nor does the route via lexicographic probability systems support $\text{nwd}_i(X_i, X_j)$ in the consequent of the PRat_{ind}-axiom, because player i would have to have a hierarchy of beliefs that all sum to one over the X_j. At w, however, player i has a hierarchy of beliefs such that the beliefs do not restrict themselves to the subgame that X_i and X_j span; every strategy in the whole game receives non-zero probability somewhere in the hierarchy. As a result, only weak undomination with respect to the original game, and not to a subgame, is correct.

2.4 Mixed Iterated Strict Weak Dominance

2.4.1 The Epistemic Characterisation Theorem

If all players of a normal form game are correctly informed about their own utilities, and it is common true belief among them that all players are approximately correctly informed about the utilities of their opponents, and, in addition, it is common true belief among the players that they are rational, then they will arrive at a mixed iterated strict weak dominance solution. This is the content of a new epistemic characterisation result presented in this section. Eddie Dekel and Drew Fudenberg were among the first to look at such forms of payoff-uncertainty, and their pioneering 1990 paper gave rise to a literature relating common knowledge of payoff-uncertainty and rationality to the iterated elimination of strictly dominated strategies preceded by one round of elimination of weakly dominated strategies—the Dekel–Fudenberg procedure.[35] A key assumption guiding this literature is that payoff-uncertainty be cast in probabilistic terms. In contrast to this literature, a non-

---

[35] Dekel and Fudenberg, art. cit. Tilman Börgers, 'Weak Dominance and Approximate Common Knowledge', *Journal of Economic Theory*, 64 (1994), 265–276 characterises the Dekel–Fudenberg procedure in terms of approximate common knowledge of rationality. Adam Brandenburger, 'Lexicographic Probabilities and Iterated Admissibility', in P. Dasgupta, et al. (eds.), *Economic Analysis of Markets and Games: Essays in Honor of Frank Hahn* (Cambridge, Mass.: MIT Press, 1992), 282–276 provides a characterisation in terms of a lexicographic version called common first-order knowledge. See also Adam Brandenburger, Amanda Friedenberg and H. Jerome Keisler, art. cit. Faruk Gul, 'Rationality and Coherent Theories of Strategic Behavior', *Journal of Economic Theory*, 70 (1996), 1–31 uses the notion of the weakest perfect τ-theory. P. Jean-Jacques Herings and Vincent Vannetelbosch, 'The Equivalence of the Dekel–Fudenberg Iterative Procedure and Weakly Perfect Rationalizability', *Economic Theory*, 15 (2000), 677–687 describe the solution concept in terms of the beliefs players may have about the possibility of opponents making errors with small, correlated probability. Elchanan Ben-Porath, 'Rationality, Nash Equilibrium and Backwards Induction in Perfect-Information Games', *Review of Economic Studies*, 64 (1997), 23–46 studies very similar payoff-uncertainty in extensive games.

probabilistic model of payoff-uncertainty is developed here, showing that under such a conception, common belief about payoff-uncertainty and rationality characterises a new solution concept of mixed iterated strict weak dominance.

If $\Gamma = (I, (A_i)_i, (u_i)_i)$ is an N-person normal form game, and $X_i \subseteq A_i$ are sets for all i, I write $(I, (X_i)_i, (u_i|_{X_i})_i)$ for the game resulting from Γ by removing for all i the strategies in the complement of X_i with respect to A_i and restricting the utility functions correspondingly, and apply the notation

$$\mathrm{nsd}_i(X_1, \ldots, X_N)$$

for the pure strategies that are not strictly dominated for player i in the subgame of Γ spanned by $\prod_i X_i$. The nwd_i are defined analogously. Using the nsd_i and the nwd_i, *iterated strict dominance*, as I have illustrated, is captured by

$$\begin{aligned} \mathsf{S}_i^0 &= A_i, \\ \mathsf{S}_i^{n+1} &= \mathrm{nsd}_i(\mathsf{S}_1^n, \ldots, \mathsf{S}_N^n) \quad (n \geq 0); \end{aligned}$$

and the concept characterised here, *mixed iterated strict weak dominance*, by

$$\begin{aligned} \mathsf{M}_i^0 &= A_i, \\ \mathsf{M}_i^{n+1} &= \mathrm{nwd}_i(\mathsf{S}_1^n, \ldots, \mathsf{M}_i^n, \ldots, \mathsf{S}_N^n) \quad (n \geq 0), \end{aligned}$$

showing mixed recursion. Informally put, the idea is that player i considers a sequence of games spanned by, for opponent strategies, the relevant stages from the sequence of iterated strict dominance, and for herself, the strategies that are not weakly dominated for her in the previous stage. Stage zero is A_i. To obtain stage one, she removes from the entire game those strategies of hers that are weakly dominated. To obtain stage two, she considers the game spanned by, for opponent strategies, the sets S_j^1, $j \neq i$, and for herself, the set obtained at stage one (that is, $\mathrm{nwd}(A_1, \ldots, A_i, \ldots, A_N)$), and she removes from this game those strategies of hers that are weakly dominated, and so she continues.

Just as before, the sets of proposition letters for these solution concepts are denoted S_i^∞ and M_i^∞, respectively, with super-scripts n indicating what remains after n rounds of elimination.

This models the outcome of game-playing. In order to model part of the antecedent conditions—payoff-uncertainty—we need to be able to talk not only about restrictions of the game, but also about games in which utility functions have been slightly altered. The following ingredients return in my axiomatisation below. If player i's strategies are written i_1, i_2, and so on, a multi-matrix $(r_{i,k_1,\ldots,k_N})_{i,k_1,\ldots,k_N}$ containing reals $r_{i,k_1,\ldots,k_N}$ is used to build constructs of the form

$$\mathrm{nsd}_i(X_1, \ldots, X_N, (r_{i,k_1,\ldots,k_N})_{i,k_1,\ldots,k_N})$$

denoting the set of pure strategies that are not strictly dominated in the subgame of Γ spanned by $\prod_i X_i$ in which the utility functions u_i are replaced by utility functions u_i' satisfying

$$u_i'(1_{k_1},\ldots,N_{k_N}) = r_{i,k_1,\ldots,k_N}.$$

That is, take Γ, remove all strategies in the complement of X_i, substitute u_i by u_i', and collect all strategies that are not strictly dominated in the resulting game.

Generalising this, a multi-matrix containing sets of reals $(D_{i,k_1,\ldots,k_N})_{i,k_1,\ldots,k_N}$ is used to build constructs of the form

$$\mathrm{nsd}_i(X_1,\ldots,X_N,(D_{i,k_1,\ldots,k_N})_{i,k_1,\ldots,k_N})$$

denoting the set of pure strategies of player i that are not strictly dominated in any subgame spanned by $\prod_i X_i$ in which the utility functions u_i are replaced by utility functions u_i' satisfying

$$u_i'(1_{k_1},\ldots,N_{k_N}) \in D_{i,k_1,\ldots,k_N}.$$

That is, take as many copies of Γ as there are u_i' satisfying this condition, remove all strategies in the complement of X_i, substitute u_i by u_i' in the corresponding copy of Γ, and collect all strategies that are not strictly dominated in any resulting game.

Dekel and Fudenberg's objective was to investigate what players of normal form games end up playing if they are less than fully informed about each others' utility function in a context in which rationality prescribes players to exclude weakly dominated strategies.[36] As I suggested in my examination of the general form of characterisation results, players cannot be fully ignorant, since otherwise nothing can be said about their behaviour, so what Dekel and Fudenberg did was to assume that the players have approximately correct beliefs about each other.

Though not disputing the mathematical validity of their model, there is room, I believe, for an alternative approach. The basic difference, to which I will turn in more detail later, is that while Dekel and Fudenberg model approximately correct beliefs in probabilistic terms, the approach presented here uses epistemic logic.

The first task is to formalise the statement that player i has approximately correct beliefs—*approximate beliefs*—about player j's utility function. I assume without loss of generality that there are two players. What makes the formalisation task a non-trivial one is that this has to be accomplished in a context in which player i also believes that player j is correctly informed—has *exact beliefs*—about her own utility function. I will first formalise player i's beliefs about player j's exact beliefs, and then turn to player i's approximate beliefs.

The most natural way to formalise the statement that player j has exact beliefs about the utility $r_{j,k,l}$ player j assigns to strategy profile (k,l) is

$$\Box_j \mathbf{u}_j(k,l) = \mathbf{r}_{j,k,l},$$

and thus a simple way to formalise the statement that player i believes that player j has exact beliefs about the utility player j assigns to strategy profile (k,l) is to add a $\Box_i$ to the above sentence, resulting in

---

[36] Dekel and Fudenberg, art. cit. 245 write that 'Each player knows his/her own payoffs, and so by our rationality postulate will not choose a weakly dominated strategy'.

$$\Box_i \Box_j \mathbf{u}_j(k,l) = \mathbf{r}_{j,k,l}.$$

Yet this cannot be coherent in a context where player i has only approximate beliefs about player j's utility function, for in such a context $\Box_i \Box_j \mathbf{u}_j(k,l) = \mathbf{r}_{j,k,l}$ implies that player i has in mind the specific utility value $r_{j,k,l}$, illegitimately suggesting that there exists a kind of exact belief about player j's utility function. What does work, though, is to cast player i's beliefs about player j's exact beliefs about the utility player j assigns to strategy profile (k,l) in terms of a belief about a conjunction of implications,

$$\Box_i \bigwedge_{r \in D_{j,k,l}} (\mathbf{u}_j(k,l) = \mathbf{r} \to \Box_j \mathbf{u}_j(k,l) = \mathbf{r}),$$

for a finite set $D_{j,k,l}$ containing $r_{j,k,l}$ and contained in a small environment of $r_{j,k,l}$. To put it informally, this expresses player i's belief that if the utility player j assigns to (k,l) takes, say, value r, then player j believes that it takes value r, and this conditional does not imply that player i has particular utility values in mind. Generalising this,

$$\Box_i \bigwedge_{k,l} \bigwedge_{r \in D_{j,k,l}} (\mathbf{u}_j(k,l) = \mathbf{r} \to \Box_j \mathbf{u}_j(k,l) = \mathbf{r})$$

expresses the fact that player i believes that player j has exact beliefs about player j's utility function. This is abbreviated by $\Box_i \Box_j \upsilon_j$, and

$$\bigwedge_{k,l} \bigwedge_{r \in D_{j,k,l}} (\mathbf{u}_j(k,l) = \mathbf{r} \to \Box_j \mathbf{u}_j(k,l) = \mathbf{r})$$

is abbreviated by $\Box_j \upsilon_j$. The fact that this notation does not fully capture the logical form of the statements is unproblematic.

The careful formalisation of beliefs about exact beliefs about a utility function makes it rather straightforward to formalise approximate beliefs about a utility function. Due to the above problem about illegitimate beliefs about specific utility values, it is not an option to represent player i's approximate beliefs about the utility player j assigns to strategy profile (k,l) by means of

$$\Box_i \mathbf{u}_j(k,l) = \mathbf{r}'$$

for some r' sufficiently close to the real $r_{j,k,l} = u(k,l)$. What does work, however, is to mark

$$\Box_i \bigvee_{r \in D_{j,k,l}} \mathbf{u}_j(k,l) = \mathbf{r}$$

as a finite set of reals $D_{j,k,l}$, in the way mentioned before, containing $r_{j,k,l}$ and contained in a small environment of $r_{j,k,l}$, and to define the degree of approximation by putting specific conditions on $D_{j,k,l}$. Dekel and Fudenberg suggest such a natural condition, which is adopted here.[37] Given a particular two-person normal form game, one can find $\varepsilon_{j,k,l} > 0$ such that if $|r - r_{j,k,l}| < \varepsilon_{j,k,l}$ for all $r \in D_{j,k,l}$, player i's

[37] Ibid. 245.

beliefs are such that player i gets relations of strict dominance among the strategies of player j right, but not necessarily relations of weak dominance. This is expressed by saying that the $\varepsilon_{j,k,l}$ *ensure correct beliefs* about strict dominance whenever these conditions hold with respect to the utility player j assigns to strategy profile (k,l).

Given the earlier discussion of iterated weak dominance, it is easy to formalise the rationality requirement that no weakly dominated strategy be played. The basis is captured by

MRat$_{bas}$ $(\mathbf{mrat}_i \wedge \Box_i \bigwedge_{i_k \in A_i, j_l \in A_j} \mathbf{u}_i(k,l) = \mathbf{r}_{i,k,l}) \rightarrow \mathrm{nwd}_i(A_i, A_j, (\mathbf{r}_{i,k,l})_{i,k,l})$;

the inductive step, by

MRat$_{ind}$ $(\mathbf{mrat}_i \wedge \Box_i \bigwedge_{i_k \in X_i, j_l \in X_j} \mathbf{u}_i(k,l) = \mathbf{r}_{i,k,l} \wedge \Box_i X_i \wedge \Box_i X_j) \rightarrow$
$\mathrm{nwd}_i(X_i, X_j, (\mathbf{r}_{i,k,l})_{i,k,l})$.

In summary, I have formalised two elements of payoff-uncertainty (beliefs about exact beliefs about utility functions and approximate beliefs about utility functions) as well as Dekel and Fudenberg's notion of rationality excluding weakly dominated strategies. In order to investigate game-playing situations with common belief about payoff-uncertainty and common belief about rationality, we have to take one more step to ensure that these two components of the antecedent do the job that they are supposed to do. It may come as a surprise that we need extra axioms to that end, as we needed nothing of that kind in the epistemic characterisation results that were dealt with earlier. There is a peculiar obstacle to applying the necessitation role, however, and for that reason we need to do more here.

To see this, observe that if we apply necessitation for j and the K-axiom to the basis axiom for player i we find

$$(\Box_j \mathbf{mrat}_i \wedge \Box_j \Box_i \bigwedge_{i_k \in A_i, j_l \in A_j} \mathbf{u}_i(k,l) = \mathbf{r}_{i,k,l}) \rightarrow \Box_j \mathrm{nwd}_i(A_i, A_j, (\mathbf{r}_{i,k,l})_{i,k,l}).$$

The antecedent of this sentence cannot, however, be made true under the assumption of common belief about payoff-uncertainty and rationality. This is because the second conjunct of the antecedent involves beliefs possessed by player j about player i's beliefs about specific utility values player i assigns to certain outcomes of the game, and as I have demonstrated, this is incoherent in a context in which player j has only approximate beliefs about player i's utility function. In fact, it is a good thing that the antecedent of the above sentence cannot be made true under the assumption of common belief about payoff-uncertainty and rationality, for if it were true, player j would have correct beliefs about weak dominance relations among player i's strategies. Such beliefs were excluded because approximate beliefs about utility functions were defined in terms of getting strict dominance relations right, but not necessarily weak dominance relations.

So, necessitation does not get the right procedure off the ground without extra axioms. The solution proposed here is to use clauses of the form

$$(\Box_j \mathbf{mrat}_i \wedge \Box_j \Box_i \upsilon_i \wedge \Box_j \bigwedge_{k,l} \bigvee_{r \in D_{i,k,l}} \mathbf{u}_i(k,l) = \mathbf{r}) \rightarrow \Box_j \mathrm{nsd}_i(A_i, A_j, (D_{i,k,l})_{i,k,l}),$$

to express the consequences of player j's beliefs about player i's rationality in a situation in which player j only has approximate beliefs about player i's utility functions. Clearly, the antecedent conditions are fulfilled once common belief about payoff-uncertainty and rationality is assumed; player j believes that player i is rational (first conjunct), player j believes that player i has exact beliefs about player i's utility function (second conjunct), and player j has approximate beliefs about player i's utility function (third conjunct). Equally without doubt, the consequent provides the appropriate beliefs, as player j believes that player i chooses a strategy that is not *strictly* dominated.

More intricate, but structurally similar reasoning occurs in the proof of the epistemic characterisation result, for which we need more intricate, but structurally similar axioms. In fact, again there is a distinction between the two cases, depending on whether player i has additional beliefs in the form of $\Box_i X_i$ and $\Box_i X_j$ or not.

Knw$_{bas}$ $(\Box_j \Box_i^n \mathbf{mrat}_i \wedge \Box_j \Box_i^n \Box_i \upsilon_i \wedge \Box_j \bigwedge_{k,l} \bigvee_{r \in D_{i,k,l}} \mathbf{u}_i(k,l) = \mathbf{r}) \rightarrow$
$\Box_j \Box_i^n \mathrm{nsd}_i(A_i, A_j, (D_{i,k,l})_{i,k,l})$.
Knw$_{ind}$ $(\Box_j \Box_i^n \mathbf{mrat}_i \wedge \Box_j \Box_i^n \Box_i \upsilon_i \wedge \Box_j \Box_i^n X_i \wedge \Box_j \Box_i^n X_j \wedge$
$\Box_j \bigwedge_{i_k, j_l} \bigvee_{r \in D_{i,k,l}} \mathbf{u}_i(k,l) = \mathbf{r}) \rightarrow \Box_j \Box_i^n \mathrm{nsd}_i(X_i, X_j, (D_{i,k,l})_{i,k,l})$.

Without loss of generality, these axioms are phrased for a two-person normal form game so as to grasp the beliefs player j has about player i's rationality and beliefs ($n = 0$), his beliefs about player i's beliefs about player i's rationality and beliefs ($n = 1$), his beliefs about player i's beliefs about player i's beliefs about player i's rationality and beliefs, and so on ($n > 1$). The finiteness of the game ensures that only finitely many are really needed.

Table 2.6

Assumptions	
preferences	$\bigwedge_{i,k,l} \mathbf{u}_i(k,l) = \mathbf{r}_{i,k,l}$
principles	$\bigwedge_i \mathbf{mrat}_i$
beliefs	
preferences	$\Box_i \bigwedge_{k,l} \mathbf{u}_i(k,l) = \mathbf{r}_{i,k,l}$
	$\Box_i \bigwedge_{j \neq i,k,l} \bigvee_{r \in D_{i,k,l}} \mathbf{u}_j(k,l) = \mathbf{r}_{j,k,l}$
	$\mathbf{C} \bigwedge_i \Box_i \upsilon_i$
	$\mathbf{C} \bigwedge_i \Box_i \bigwedge_{j \neq i,k,l} \bigvee_{r \in D_{i,k,l}} \mathbf{u}_j(k,l) = \mathbf{r}_{j,k,l}$
principles	$\mathbf{C} \bigwedge_i \mathbf{mrat}_i$
performed action	–
Solution Concept	
player i	M_i^∞
Proof System	$_{\Gamma}\mathbf{K4_{EC}mrat}$

A proof system $_{\Gamma}\mathbf{K4_{EC}mrat}$ is used consisting of Prop, Dual, K, 4, E, C, the proof rules modus ponens, necessitation and induction, the first three axioms for normal form game-playing situations (KnUt is omitted), plus four MRat- and Knw-axioms. The assumptions that underlie the characterisation result are routinely listed in Table 2.6.

Theorem 2.6 *Let $\Gamma = (I, (A_i)_i, (u_i)_i)$ be a two-person normal form game, let $i = 1, 2$, $j = 3 - i$, and let $D_{i,k,l}$ be finite sets of reals such that $|r - u_{i,k,l}| < \varepsilon_{i,k,l}$ for all $r \in D_{i,k,l}$ and $\varepsilon_{i,k,l}$ ensuring correct beliefs about strict dominance. Assume that the following three conditions are true.*

1. *All players have true beliefs concerning their own utilities.*
2. *It is common true belief among the players that they have approximate beliefs about the utility functions of their opponents.*
3. *It is common true belief among the players that they are rational.*

Then the players play strategies that survive the mixed iterated strict weak dominance elimination procedure.

Proof I write φ^n for

$$\bigwedge_i \mathbf{mrat}_i \wedge \bigwedge_{i,k,l} \mathbf{u}_i(k,l) = \mathbf{r}_{i,k,l} \wedge \bigwedge_i \Box_i \bigwedge_{k,l} \mathbf{u}_i(k,l) = \mathbf{r}_{i,k,l} \wedge$$

$$\Box_i \bigwedge_{k,l} \bigvee_{r \in D_{j,k,l}} \mathbf{u}_j(k,l) = \mathbf{r} \wedge \mathbf{E}^{\leq n} \Box_i \upsilon_i \wedge \mathbf{E}^{\leq n} \Box_i \bigwedge_{k,l} \bigvee_{r \in D_{j,k,l}} \mathbf{u}_j(k,l) = \mathbf{r} \wedge \mathbf{E}^{\leq n} \bigwedge_i \mathbf{mrat}_i$$

and φ_i^n for the part of φ^n starting with $\Box_i$ conjoined with the statement $\mathbf{mrat}_i$, that is,

$$\mathbf{mrat}_i \wedge \Box_i \bigwedge_{k,l} \mathbf{u}_i(k,l) = \mathbf{r}_{i,k,l} \wedge \Box_i \bigwedge_{k,l} \bigvee_{r \in D_{j,k,l}} \mathbf{u}_j(k,l) = \mathbf{r} \wedge$$

$$\Box_i \mathbf{E}^{\leq n-1} \Box_i \upsilon_i \wedge \Box_i \mathbf{E}^{\leq n-1} \Box_i \bigwedge_{k,l} \bigvee_{r \in D_{j,k,l}} \mathbf{u}_j(k,l) = \mathbf{r} \wedge \Box_i \mathbf{E}^{\leq n-1} \bigwedge_i \mathbf{mrat}_i,$$

with the convention that if $n = 0$ the $\Box_i \mathbf{E}^{\leq n-1}$ vanish completely. It has to be shown that

$$\forall n \vdash \varphi_i^n \to M_i^{n+1}. \tag{2.2}$$

First, however, it is proven that all you need to prove 2.2 is

$$\forall n \vdash \varphi_i^n \to (\Box_i M_i^n \wedge \Box_i S_j^n). \tag{2.3}$$

Assume, indeed, that 2.3 is proven. By definition of φ_i^n we also have

$$\forall n \vdash \varphi_i^n \to (\mathbf{mrat}_i \wedge \Box_i \bigwedge_{k,l} \mathbf{u}_i(k,l) = \mathbf{r}_{i,k,l}).$$

Applying MRat$_{ind}$ for i yields

$$\forall n \vdash \varphi_i^n \to \mathrm{nwd}_i(M_i^n, S_j^n),$$

which, observing that $M_i^{n+1} = \text{nwd}_i(M_i^n, S_j^n)$, concludes the proof the statement.

One level deeper, all you need to prove 2.3 is

$$\forall n \vdash \varphi_i^n \to (\Box_i\Box_i M_i^{n-1} \wedge \Box_i\Box_i S_j^{n-1} \wedge \Box_i\Box_j S_j^{n-1} \wedge \Box_i\Box_j S_i^{n-1}). \tag{2.4}$$

For assume that (2.4) is proven. To show

$$\forall n \vdash \varphi_i^n \to \Box_i M_i^n,$$

by definition of φ_i^n and the assumption we have

$$\forall n \vdash \varphi_i^n \to (\Box_i \mathbf{mrat}_i \wedge \Box_i\Box_i \bigwedge_{k,l} \mathbf{u}_i(k,l) = \mathbf{r}_{i,k,l} \wedge \Box_i\Box_i M_i^{n-1} \wedge \Box_i\Box_i S_j^{n-1}).$$

Apply the rule of necessitation for i to the appropriate instance of MRat$_{ind}$ for i, and observe that its consequent is what had to be shown and that its antecedent is what was just shown. To show

$$\forall n \vdash \varphi_i^n \to \Box_i S_j^n,$$

by definition of φ_i^n and the assumption we have

$$\forall n \vdash \varphi_i^n \to \Box_i \mathbf{mrat}_j \wedge \Box_i\Box_j \upsilon_j \wedge \Box_i\Box_j S_j^{n-1} \wedge \Box_i\Box_j S_i^{n-1} \wedge \Box_i \bigwedge_{k,l} \bigvee_{r \in D_{j,k,l}} \mathbf{u}_j(k,l) = \mathbf{r}.$$

Take the appropriate instance of Knw$_{ind}$ for i (the instance of the axiom that speaks about i's beliefs about j's iterated beliefs), and observe that its consequent is what had to be shown and that its antecedent is what was just shown. This concludes the proof that all you need to prove (2.3) is (2.4).

It remains to be shown that (2.4) is true. Induction on n. The basis is left to the reader. Assume that the statement holds for some n. I prove four implications for $n+1$ separately.

To prove

$$\varphi_i^{n+1} \to \Box_i\Box_i M_i^n,$$

it is sufficient to prove $\varphi_i^{n+1} \to (\Box_i\Box_i\Box_i M_i^{n-1} \wedge \Box_i\Box_i\Box_i S_j^{n-1})$. Assume the latter to be true. Apply the rule of necessitation for i to the appropriate instance of MRat$_{ind}$ for i twice, and observe that its consequent is what had to be shown and that its antecedent follows from the definition of φ_i^{n+1} and the assumption.

To prove the truth of the assumption apply necessitation for i to the inductive hypothesis and observe that $\forall n \vdash \varphi_i^{n+1} \to \Box_i \varphi_i^n$.

To prove

$$\varphi_i^{n+1} \to \Box_i\Box_i S_j^n,$$

it is sufficient to prove $\varphi_i^{n+1} \to (\Box_i\Box_i\Box_j S_j^{n-1} \wedge \Box_i\Box_i\Box_j S_i^{n-1})$. Assume the latter to be true. Apply the rule of necessitation for i once to the appropriate instance of Knw$_{ind}$ for i and for $n = 0$, and observe that its antecedent follows from the definition of φ_i^{n+1} and the assumption, and that the consequent is what had to be shown.

To prove the truth of the assumption apply necessitation for i to the inductive hypothesis and observe that $\forall n \vdash \varphi^{n+1} \to \Box_i \varphi_i^n$.

To prove

$$\varphi_i^{n+1} \to \Box_i \Box_j S_j^n,$$

it is sufficient to prove $\varphi_i^{n+1} \to (\Box_i \Box_j \Box_j S_j^{n-1} \wedge \Box_i \Box_j \Box_j S_i^{n-1})$. Assume the latter to be true. Take the appropriate instance of Knw_{ind} for i and for $n = 1$, and observe that its consequent follows from the definition of φ_i^{n+1} and the assumption.

To prove the truth of the assumption observe that, first, $\forall n \vdash \varphi_i^{n+1} \to \varphi_i^n$, and second, $\forall n \vdash (\Box_i \Box_j S_j^{n-1} \wedge \Box_i \Box_j S_i^{n-1}) \to (\Box_i \Box_j \Box_j S_j^{n-1} \wedge \Box_i \Box_j \Box_j S_i^{n-1})$. Together with the inductive hypothesis this shows the truth of the assumption.

To prove

$$\varphi_i^{n+1} \to \Box_i \Box_j S_i^n,$$

finally, it is sufficient to prove $\varphi_i^{n+1} \to (\Box_i \Box_j \Box_i S_i^{n-1} \wedge \Box_i \Box_j \Box_i S_j^{n-1})$. Assume the latter to be true. Apply necessitation for i to the appropriate instance of Knw_{ind} for j and for $n = 0$, and observe that its antecedent follows from the definition of φ_i^{n+1} and the assumption and that its consequent is what had to be shown.

To prove the truth of the assumption, the inductive hypothesis may be assumed to hold for φ_j^n as well. Applying the rule of necessitation for i to it yields $\Box_i \varphi_j^n \to (\Box_i \Box_j \Box_i S_i^{n-1} \wedge \Box_i \Box_j \Box_i S_j^{n-1})$. What we now need is $\varphi_i^{n+1} \to \Box_i \varphi_j^n$. That is easy to see, though.

2.4.2 Discussion

The motivation for proving epistemic characterisation theorems in the Epistemic Programme typically comes from one of two sides. One may start with a familiar game-theoretic solution concept in mind, and ask under what epistemic conditions it will capture game-play adequately; or one may start with certain epistemic conditions in mind, and investigate what solution concept follows—the *behavioural consequences* of certain epistemic conditions. Taking the first route may lead to the discovery of new and surprising epistemic conditions; taking the second route, to new and surprising solution concepts.

While Dekel and Fudenberg started from the epistemic conditions of payoff-uncertainty, and linked them to what would become known as the Dekel–Fudenberg procedure (the second route), most of the ensuing game-theoretic literature remained with the Dekel–Fudenberg procedure to study modifications of Dekel and Fudenberg's epistemic conditions (the first route). In contrast to this literature, I will begin with the epistemic conditions in mind, providing an alternative formalisation of common true belief about payoff-uncertainty and rationality to examine its behavioural consequences (the second route). The consequences uncovered by my characterisation theorem are quite different from those of the Dekel–Fudenberg procedure, however, and this is due to the fact that the formalisation of payoff-uncertainty proposed here is different. In order to locate these differences with precision, I will discuss the conceptions of payoff-uncertainty developed by Dekel and Fudenberg, as well as one by Tilman Börgers (which is fairly representative of the literature) to show by means of an example to what extent they are different from the model proposed here. I will then turn to a conceptual motivation of the axioms that connects them to the more standard, semantic tools used in the Epistemic Programme of lexicographic beliefs.

2.4.2.1 Comparison with the Literature

What Dekel and Fudenberg do is model payoff-uncertainty by means of *elaborations* of games as they were developed by John Harsanyi.[38] Roughly, a sequence of games is considered in which the utility functions of the players are slightly different from those in the original game. They define a notion of convergence, both on sequences of games and on sequences of strategies, and the main result is that a strategy survives the iterated elimination of strictly dominated strategies preceded by one round of elimination of weakly dominated strategies just in case it is the limit of a sequence of strategies that survive the iterated elimination of weakly dominated strategies in elaborations converging to the original game. Börgers, in turn, defines a notion of *approximate common knowledge*, and shows that approximate common knowledge of rationality characterises the Dekel–Fudenberg procedure.[39]

The difference between these two models and the model I propose becomes clear when we compare the respective formalisations of statements such as

> Player i has approximately correct beliefs about the utility player j assigns to strategy profile (k,l).

Dekel and Fudenberg as well as Börgers model this by having player i assign high probability to the correct, actual utility value, and low probability to alternative utility values. Under this reading, approximate beliefs are modelled in probabilistic terms. In the model I present, by contrast, player i possesses a belief concerning a disjunction of statements one of which expresses the correct utility, and all of which fall in a small environment of the correct utility. The non-probabilistic model proffered here has, one could say, player i assign equal probability to finitely many alternatives close to the actual value.

In order to demonstrate that this makes a difference, consider the games shown in Figure 2.6. The game on the left correctly represents the utility functions of both players; it is the game they actually play. The game on the right is the game as player 2 perceives it. It represents player 2's utility correctly, but it gives an approximation of the utility player 1 assigns to strategy profile $(1_1, 2_2)$. The intended interpretation

[38] 'Games with Incomplete Information Played by "Bayesian" Players' (three parts), *Management Science*, 14 (1967–1968), 159–182, 320–334, 486–502.

[39] Art. cit. A proposition φ is approximate common knowledge whenever everyone believes with high probability that φ, everybody believes with high probability that everyone believes with high probability that φ, etc.

	2_1	2_2	2_3
1_1	(1,4)	(1,8)	(1,0)
1_2	(5,8)	(1,4)	(3,0)
1_3	(3,4)	(0,8)	(2,0)

	2_1	2_2	2_3
1_1	(1,4)	(D,8)	(1,0)
1_2	(5,8)	(1,4)	(3,0)
1_3	(3,4)	(0,8)	(2,0)

Fig. 2.6

is that player 2 believes that $u_1(1_1, 2_2) \in D$ for some finite set D containing the actual utility value and contained in a small environment of it, such that the conditions of the epistemic characterisation theorem are satisfied. That player 2 has approximate beliefs about the utility player 1 assigns to $(1_1, 2_2)$ is therefore formalised here as a belief about a disjunction over D. Strictly speaking, the game on the right represents not one but $|D|$ games. For completeness, I should mention that player 1 perceives the game as it is. The only outcome that survives the Dekel–Fudenberg elimination procedure is $(1_2, 2_1)$. If player 2 assigns high probability p to the correct utility player 1 assigns to $(1_1, 2_2)$, any other outcome disappears when p tends to infinity. Similar observations hold for Börgers' model. If, however, we assume common belief about payoff-uncertainty and rationality as formalised here, then there is a second possible outcome: $(1_2, 2_2)$. An informal way to demonstrate this as follows. First note that under common true belief about payoff-uncertainty and rationality, player 1 will play her second strategy, because both her first one and her third one are weakly dominated. To see that player 2 can play his second strategy under these assumptions, observe that, since he is rational, he does not play the weakly dominated 2_3. Furthermore, player 2 believes that player 1 is rational. From the game which player 2 thinks is being played (the matrix on the right), player 2 removes player 1's weakly dominated strategies, but the only such strategy is 1_3. Particularly, since player 2 only has approximate beliefs about the utility function of player 1, he cannot justifiably remove 1_1: player 2 holds it possible that 1_1 will score more for player 1 against 2_2 than any other strategy of player 1. From player 2's point of view, player 1 will either play her first or her second strategy, and this makes, for player 2, playing his first as well as his second strategy rational. Rationality does not exclude 2_2, and hence $(1_2, 2_2)$ is a possible outcome.

2.4.2.2 Motivation of the Axioms

While the syntactic approach developed here makes it possible to analyse the role of various levels of common belief about rationality and a particular form of payoff-uncertainty, it has the drawback of making it more difficult to motivate the axioms in terms of the lexicographic beliefs standardly used in the Epistemic Programme in the theory of games. While to some extent this is a feature of syntactic investigations, such a motivation may justifiably be demanded, and I will attempt to provide one here.

To start with, the purpose of the MRat_{bas}-axiom is to capture the rationality principle according to which weakly dominated strategies are to be excluded, and this is accomplished by stating that player i will not play a strategy that is weakly dominated in the entire game on the basis of no more information than that player i is rational and that she is correctly informed about her own utility function. A high-yielding place to look for a basis of a motivation for this axiom is a lemma encountered earlier—Lemma 2.3, according to which a strategy is not weakly dominated for player i in the entire game whenever it is a best response for her to some probability measure with full support (that is, assigning zero probability to no alternative) over player j's available strategies. As I suggested in the discussion of iterated weak dominance, however, some care has to be taken not to misuse the full-support requirement.

MRat_{bas} indeed formalises a situation in which no specific information is available about which alternatives player i's beliefs exclude. In the process of iteratively eliminating strategies, player i's beliefs focus on certain sets of strategies of player j and herself. By doing so her beliefs start excluding alternatives, and accordingly cease to have full support (the temporal phrasing of this process is rather metaphoric). At the first stage of the elimination process player i assigns zero probability to no strategy of her opponent, but at the next stage of the elimination process she removes strategies that, given her own payoff-uncertainty and beliefs about her opponent j's rationality, she believes j not to play. If such strategies exist, player i's beliefs at this stage exclude them and consequently do not have full support.

In my formalism, MRat_{ind} deals with this case. In order to motivate this axiom, consider the following first attempt. If player i's beliefs are captured by $\Box_i X_i \wedge \Box_i X_j$ for $X_i \subseteq A_i$ and a real subset $X_j \subset A_j$, they have indeed no full support with respect to the entire game, but they do have full support with respect to the subgame spanned by X_i and X_j. This shows, the attempt would go, that Lemma 2.3 may still be used to justify why MRat_{ind} has player i choose a strategy that is not weakly dominated in that very subgame—simply reduce it to the subgame. This may sound like a plausible defence of MRat_{ind}, but it has an important defect. Going from the first to the second stage, player i has given up her full-support beliefs from the first stage. While these beliefs have full support with respect to the subgame of the second stage, the fact that they no longer have full support with respect to the entire game makes it impossible simultaneously to adopt MRat_{bas} (which seems to presuppose full-support beliefs with respect to the entire game) and MRat_{ind} (which seems to presuppose non-full-support beliefs).

This is very much related to a puzzle presented by Larry Samuelson.[40] A solution to that puzzle by means of lexicographic probability systems by Adam Brandenburger, Amanda Friedenberg and H. Jerome Keisler proves very useful for the mo-

40 'Dominated Strategies and Common Knowledge', *Games and Economic Behavior*, 4 (1992), 284–313.

	2_1	2_2
1_1	(1,1)	(0,1)
1_2	(0,2)	(1,0)

Fig. 2.7

tivational task.[41] Consider their version of Samuelson's game shown in Figure 2.7. If player 2 is rational in Dekel and Fudenberg's sense of not choosing weakly dominated strategies, he will not play 2_2. This means that if player 1 expects player 2 to be rational in that sense, she will not form a full-support belief. But how is that possible if at the same time player 1 also rationally excludes weakly dominated strategies? There seems to be a friction between assuming someone to be rational in the sense of Dekel and Fudenberg, and assuming someone to believe others to be rational in that same sense.

Brandenburger, Friedenberg and Keisler's solution represents the players' beliefs via lexicographic probability systems. Player 1's primary measure assigns probability 1 to player 2 playing 2_1, thus grasping the consequences of her expectations about player 2's rationality. But player 1 has a secondary measure assigning probability 1 to the event that player 2 plays 2_2, and the interpretation of this is that player 1 considers it infinitely more likely that player 2 is rational and plays 2_1 than that he plays 2_2. The idea now is that player 1's beliefs at the end of the iterative elimination process that the three authors consider (which differs from ours) can be represented by means of a lexicographic probability system the primary measure of which assigns positive probability to the surviving strategies only, while the remaining measures cover the subsequently eliminated strategies. In doing so, these authors rely on the fact that a strategy is not weakly dominated for player i in the entire game nor in the subgame spanned by X_i and X_j whenever it is a best response to a lexicographic probability system with full support (all alternatives receive nonzero probability in some measure in the system) according to which X_i is infinitely more likely than its complement. The situation described by this result is exactly that covered by MRat_{ind} without any resulting incoherence with MRat_{bas}.

This motivates MRat_{bas} and MRat_{ind} in terms of lexicographic beliefs. Knw_{bas} and Knw_{ind}, in turn, are intended to capture the specific conception of payoff-uncertainty: if player j is fully informed about player i's rationality as well as about the fact that player i has correct beliefs about her own utility function, but at the same time player j is less than fully informed about player i's utility function, then player j will form certain beliefs about player i's prospective strategies as well as about player i's beliefs about her own prospective strategies.

[41] Art. cit.

At first sight there may seem to be an incoherence—if not an inconsistency—as the antecedent rationality condition of Knw_{bas} and Knw_{ind} refers to a rationality concept that, by MRat_{bas} and MRat_{ind}, was cast in terms of excluding weakly dominated strategies, while the consequent refers to the exclusion of strong domination. It is important to realise, however, that in Knw_{bas} as well as in Knw_{ind} strong domination does not occur in the sense of an alternative to weak domination, and that it does not entail that player j would attribute to player i something like the possession of lexicographic (or non-lexicographic) beliefs. Rather, its occurrence is motivated by the fact that player j is under the sway of a particular form of payoff-uncertainty. She is so cautious that she is unwilling to exclude—epistemically—those strategies that are weakly dominated but not strictly dominated. An additional sign of her cautious epistemic attitude is the requirement, in the consequent of Knw_{bas} and Knw_{ind}, that these strategies are strictly dominated in *all* games that player i holds epistemically open. Player j may reason that she does not have any information about the likelihood of the games captured by the sets $D_{i,k,l}$ (in crucial contrast, as mentioned earlier, to probabilistic approaches to payoff-uncertainty and the approach proffered in this book). Being unwilling to take any epistemic risk, she therefore focuses on strategies that are not strictly dominated in any of the games that she cannot epistemically exclude. This epistemic policy is motivated by combining the fact that she is epistemically risk-averse with her beliefs about player i's rationality (in the sense of excluding weakly dominated strategies). If, for instance, she knew j to be irrational and to play completely arbitrarily, she would not adopt any such specific policy and would reside with full-support equiprobable beliefs.

In summary, the assumptions of the epistemic characterisation theorem that are proven here involve players who adopt, on the level of the practical rationality of strategy choice, a principle that is broad in the sense that it allows for more than its competitor, based on strong dominance. On the level of the theoretic rationality of belief formation, by contrast, the players adopt a more limited and cautious policy.

2.4.2.3 The Ban on Exogenous Information

While this motivates the axioms and makes them intelligible in terms of lexicographic beliefs and epistemic policies, it does not yet directly show them to be plausible. I do believe, however, that nothing speaks conceptually against the assumptions the theorem makes. In particular, there seems to be ample room in the Epistemic Programme for an analysis of the consequences of interpreting payoff-uncertainty in equiprobable terms in a context of cautious belief formation policies, rather than as involving players whose uncertainty entails assigning different degrees of likelihood to alternatives.

Yet payoff-uncertainty does not seem to sit easily with the general format of the Epistemic Programme and its ban on exogenous information. Uncertainty about the utility functions of opponents does not seem to be attributable to causes internal to the game, but rather by making reference to exogenous factors. That I do not have exact beliefs about, say, the utility you assign to selling your house for a certain price is due to such exogenous factors as that I do not know you very well and that I have not listened in on your conversations with your real estate agent. It does not have any endogenous component, or so it seems.

The ban on exogenous information, however, is not a ban on exogeneity in general. The actions available to you and your opponents, for instance, have a decidedly exogenous character in that they can be causally linked to factors outside the game, and this is even true for beliefs. Even if it were the case that common belief about rationality and utility arises as a causal consequence of clearly circumscribed empirical factors (say, the administering of certain medications to the players), this would not mean that common knowledge of rationality goes against the ban on exogenous information. Similarly, payoff-uncertainty may have exogenous empirical causes without contravening the ban.

Of course, genuine exogenous payoff-uncertainty is possible. If I have seen you behave in ways that look hesitant to me, I may conclude that you are a ditherer, but I may as well conclude that I simply do not have enough information to determine your utility function very precisely. If I settle on payoff-uncertainty, I do so on exogenous grounds. To underscore the conclusion from the previous paragraph, it must be noted that the form of payoff-uncertainty encountered in the epistemic characterisation theorem is different from that in the ditherer's case, because in the characterisation result the uncertainty is systematically linked to the game structure itself by way of the condition that the $\varepsilon_{j,k,l}$ ensure correct beliefs about strict dominance. That condition is phrased in terms of the game, rather than exogenous factors, and thus excludes genuinely exogenous payoff-uncertainty.

Chapter 3
Extensive Games

Game theorists may disagree about the suitability of one or another normal form solution concept—there are many—but a brief inspection of the literature will show that they rarely disagree about the correct epistemic characterisation of each solution concept. For extensive games, while several solution concepts are available, the most obvious candidate is certainly in most cases backward induction (the subgame-perfect (Nash) equilibrium). Ironically, however, game theorists widely disagree about its correct epistemic characterisation, and the disagreement centres on the question of whether or not common true belief about rationality leads to backward induction. Robert Aumann defends a position in favour of this implication, Philip Reny objects, and both positions are taken by various other theorists.[1]

The purpose of this chapter is not to substantiate one line of argument or another. Rather, by analysing the logical form of the arguments *à la* Aumann and *à la* Reny, I will point out that important modelling assumptions have been overlooked. Devising a logical formalism that captures the two different interpretations of what extensive games in fact model, I will argue that the argument *à la* Aumann assumes the one-shot interpretation in combination with a principle of rationality with which it is incompatible, and I will argue that the argument *à la* Reny assumes the many-moment interpretation in combination with implausible belief revision policies.[2]

---

[1] Robert Aumann, 'Backward Induction and Common Knowledge of Rationality', *Games and Economic Behavior*, 8 (1995), 6–19, ibid., 'Reply to Binmore', *Games and Economic Behavior*, 17 (1996), 138–146, and ibid., 'On the Centipede Game', *Games and Economic Behavior*, 23 (1998), 97–105. Philip Reny, 'Two Papers on the Theory of Strategic Behaviour', Ph.D. diss. (Princeton University, 1988), 'Common Knowledge and Games with Perfect Information', in A. Fine and J. Leplin (eds.), *Proceedings of the Philosophy of Science Association: Volume 2* (East Lansing, Mich.: Philosophy of Science Association, 1989), 363–369, ibid., 'Common Belief and the Theory of Games with Perfect Information', *Journal of Economic Theory*, 59 (1993), 257–274, and ibid., 'Rational Behaviour in Extensive-Form Games', *Canadian Journal of Economics*, 28 (1995), 1–16.

[2] Joseph Halpern, 'Substantive Rationality and Backward Induction', *Games and Economic Behavior*, 37 (2001), 321–339 compares Aumann's approach to backward induction with Stalnaker's game models. In contrast to the comparison in the current chapter, Halpern does not distinguish

B. de Bruin, *Explaining Games*, Synthese Library 346,
DOI 10.1007/978-1-4020-9906-9_3,

3.1 The One-Shot Interpretation

3.1.1 The Epistemic Characterisation Result

Common true belief about rationality and utility, in extensive form games, yields backward induction, or so Robert Aumann and others claim.[3] To understand this result, define the *normal form* $\mathrm{nf}(\Gamma)$ of an extensive game Γ as a triple $(I,(A_i)_i,(v_i)_i)$ where I collects the players of Γ, A_i all strategies player i has in Γ, and $v_i\colon \prod_i A_i \to \mathbb{R}$ are utility functions such that

$$v_i(1_{k_1},\ldots,i_{k_i},\ldots,N_{k_N}) = u_i(O(1_{k_1},\ldots,i_{k_i},\ldots,N_{k_N})),$$

where O is a function mapping a tuple of strategies to the terminal node of the extensive game that is reached when the players play according to these strategies. I write $u_i(k,l)$ for $v_i(O(k,l))$. If $\mathrm{nf}(\Gamma) = (I,(A_i)_i,(v_i)_i)$, and if X_1, X_2, and so on, are sets of strategies satisfying $X_i \subseteq A_i$ for all i, then the *subspan* of $\mathrm{nf}(\Gamma)$ with respect to $\prod_i X_i$ is the triple $(I,(X_i)_i,(v_i|_{X_i})_i)$ obtained from $\mathrm{nf}(\Gamma)$ by removing for all i the strategies in the complement of X_i (with respect to A_i) and modifying the utility functions correspondingly.

In line with the previous chapter, notation $\mathrm{nsd}_i(X_1,\ldots,X_N)$ is extended to extensive game-playing settings for the strategies that are not strictly dominated for player i in the subspan of $\mathrm{nf}(\Gamma)$ with respect to $\prod_i X_i$; that is, strategies for which there is no strategy in X_i which does strictly better against any combination of opponents' strategies. In general, I am interested in dominance relations in subspans of the normal form of subgames of some underlying game generated by certain decision nodes; that is, in constructs of the form $\mathrm{nsd}_i^x(X_1,\ldots,X_N)$, where $X_i \subseteq A_i$, for some decision node x.

To compute such sets—the idea is simpler than the construction—consider the subgame Γ_x of Γ generated by some decision node x, construct its normal form $\mathrm{nf}(\Gamma_x)$, delete from $\mathrm{nf}(\Gamma_x)$, for all j, the strategies not coinciding on Γ_x with any strategy from X_j, find out which of the remaining strategies in the resulting subspan of $\mathrm{nf}(\Gamma_x)$ are strictly undominated, and then take all strategies from A_i coinciding on

between one-shot and many-moment interpretations of extensive game-play, nor does he give an inductive and implicit axiomatisation of rationality.

[3] 'Backward Induction and Common Knowledge of Rationality' contains a defence of this claim, based on the one-shot interpretation. Assuming a many-moment perspective, Aumann defended the same claim for smaller classes of extensive games in 'On the Centipede Game', a result discovered independently by Wlodek Rabinowicz, 'Grappling with the Centipede: Defence of Backward Induction for BI-terminating Games', *Economics and Philosophy*, 14 (1999), 95–126, John Broome and Wlodek Rabinowicz, 'Backwards Induction in the Centipede Game', *Analysis*, 59 (1999), 237–242. Magnus Jiborn and Wlodek Rabinowicz, 'Backward Induction without Full Trust in Rationality', in W. Rabinowicz (ed.), *Value and Choice: Some Common Themes in Decision Theory and Moral Philosophy: Volume 2* (Lund: Lund Philosophy Reports, 2001), 101–120 prove a many-moment characterisation for the Centipede on the basis of sufficiently strong, but not necessarily full beliefs about rationality.

Γ_x with such a strictly undominated strategy. For weak dominance, define nwd_i and its relativisations similarly.

The set containing the strategies that coincide with the backward induction strategy on the subgame generated by x is written BI_i^x, and in the logic the set of corresponding proposition letters is written BI_i^x. Assuming that the extensive game is *generic* in the sense that no player is indifferent between any two terminal nodes, BI_i^x contains all strategies prescribing the uniquely optimal action at x if x is an immediate predecessor of a terminal node. Reasoning back to the root of the game, BI_i^z collects all strategies prescribing the uniquely optimal action at decision node z under the assumption that at decision nodes $y \succ z$ higher up in the game tree all players j plays a strategy from BI^j. For the root ρ of the game, BI_i^ρ is a singleton also written BI_i. For terminal nodes x the convention is applied that $\mathsf{BI}_i^x = A_i$.

As I have suggested, while the mathematical differences between normal form games and extensive games strongly suggest that the former model situations of simultaneous and independent choice and the latter model temporally extended situations of sequential choice, we are by no means obliged to make such an interpretation. The one-shot interpretation, in fact, holds to the view that playing an extensive game is playing its normal form; that is, whenever players play an extensive game, what they actually do is choose, at one point in time, their strategies for the entire game. This does not completely determine a unique one-shot interpretation, and hence there is room for difference of opinion about the relevant kind of rationality principles. One can invoke some aspects of the sequential structure of the game to ascertain whether a strategy is rational or not, for while a strategy maps all decision nodes of a player to actions, choosing, for some decision node, one action over another implies that certain decision nodes will not be reached—the terminology is admittedly inappropriate in a one-shot context—and whether these unreached nodes are relevant or not determines different notions of rationality.

To capture the differences, I call a strategy *on-path rational* whenever the rationality depends only on what happens on the actual path through the extensive game, and I call it *off-path rational* just in case it prescribes rational actions at every decision node, reached or unreached. Aumann, for instance, asserts that

> each player chooses a *strategy*, in the usual game-theoretic sense of the term...; that is, he decides what to do at each of his vertices x in the game tree, whether or not x is reached.[4]

This seems to demonstrate that he adopts a one-shot conception of extensive gameplay. Yet he also writes that 'when deciding what to do at x, the player considers the situation *from that point on:* he acts *as if* x is reached', to conclude that

> it is this feature that distinguishes the current analysis from a strategic [i.e., normal] form analysis.[5]

[4] 'Backward Induction and Common Knowledge of Rationality', 7 (notation changed). As I am more interested here in what Aumann wishes to model, than in the resulting model itself, the stress lies on his verbal statements rather than on his formalism. This also applies to my treatment of Reny in the next section.

[5] Ibid. (emphasis in original, notation changed).

Attributing the one-shot interpretation to Aumann and also accepting this conclusion, there seems to be a difference between the one-shot interpretation and playing the normal form of an extensive game, and this is, in fact, more or less how Aumann seems to view it. He does accept the main tenet of the one-shot interpretation that the objects of choice of an extensive game are the strategies of its normal form, but he contrasts his view with 'strategic form analysis'. The reason for this is that he holds the view that in order to evaluate the rationality of a strategy, one has to go beyond the information of the normal form and inspect the prescriptions of the strategy at all decision nodes of the underlying extensive game; that is, Aumann adopts a one-shot interpretation of extensive game-play with an off-path conception of rationality. This is underscored by the statement that a rational player,

> no matter where he finds himself—at which vertex—[,]... will not knowingly continue with a strategy that yields him less than he could have gotten with a different strategy,[6]

as well as by the remark that

> for each of his vertices *x* and strategies *k*, it is not the case that [player] *i* knows that *k* would yield him a higher conditional payoff at *x* than the strategy he chooses.[7]

All in all, Aumann adopts a one-shot interpretation with off-path rationality. This is the same as playing normal form games as far as the objects of choice are concerned, but it is different with respect to the rationality principle. This is a view that I discarded in Chapter 1, but it will return in the discussion later on.

There is a problem, though. In some sense, the phrase about players who continue 'not knowingly' suggests many-moment game-playing situations. Instead of taking Aumann to put forward an incongruent claim here, I take knowledge to refer to beliefs that, in one-shot game-playing situations, players have about the moves that the full strategies of their opponents prescribe in (certain) subgames. The last quotation means, in that case, that no rational player will choose a full strategy that prescribes suboptimal moves at some decision node given the beliefs that the player has, in that one-shot game-playing situation, about what her opponents' choices of full strategies will prescribe in the subgame generated by that decision node.

3.1.1.1 An Explicit Formalisation of Rationality

I will present a direct formalisation to make this precise. While it has the advantage of staying close to the sources, it makes the proof of the epistemic characterisation theorem cumbersome, and dependent on heavy logical axioms. I will therefore subsequently turn to an alternative formalisation.

Following the above quotation quite literally, we have

$$\bigwedge_x \bigwedge_k \neg \bigvee_l \bigvee_m (\Box_i \mathbf{i}_k \wedge \Box_i \mathbf{j}_l \wedge \mathbf{u}_i^x(k,l) < \mathbf{u}_i^x(m,l)),$$

[6] Ibid.

[7] Ibid. 10 (notation changed).

which says that if player i believes that her opponent is playing his lth strategy, and i herself believes she is playing her kth strategy (and she is really playing her kth strategy), then there is no better strategy m than her kth strategy. This is equivalent with

$$\bigwedge_x \bigwedge_k \bigwedge_l \bigwedge_m \neg(\Box_i \mathbf{i}_k \wedge \Box_i \mathbf{j}_l \wedge \mathbf{u}_i^x(k,l) < \mathbf{u}_i^x(m,l)),$$

and with

$$\bigwedge_x \bigwedge_k \bigwedge_l ((\Box_i \mathbf{i}_k \wedge \Box_i \mathbf{j}_l) \rightarrow \bigwedge_m (\mathbf{u}_i^x(k,l) \geq \mathbf{u}_i^x(m,l))),$$

which brings out nicely the similarity with other rationality notions. This motivates the following two axioms as renderings of Aumann's notion of rationality

ANRat $\quad \mathbf{anrat}_i^x \leftrightarrow \bigwedge_k \bigwedge_l ((\Box_i \mathbf{i}_k^x \wedge \Box_i \mathbf{j}_l^x) \rightarrow \bigwedge_m (\mathbf{u}_i^x(k,l) \geq \mathbf{u}_i^x(m,l)).$
AFRat $\quad \mathbf{afrat}_i \leftrightarrow \bigwedge_{\rho \preceq x} \mathbf{anrat}_i^x.$

To prove the epistemic characterisation theorem, the proof system $_{\Gamma}\mathbf{KT_{EC}45afrat}$ is used containing Prop, Dual, K, T, 4, 5, E, C, the proof rules modus ponens, necessitation and induction, all axioms for one-shot game-playing situations, plus the two rationality axioms.

Theorem 3.1 (Aumann, 1995) *Let Γ be a finite N-person generic extensive form game with perfect information. Assume that the following two conditions are true.*

1. *There is common true belief among the players about the utility functions of all players.*
2. *It is common true belief among the players that they are rational.*

Then the backward induction outcome is reached.

To give some impression of how this theorem is proven, it can be demonstrated that for all players i we have

$$\forall x \vdash \mathbf{Cafrat} \rightarrow BI_i^x.$$

The rule of necessitation, the K-axiom, and some propositional logic and some aggregation of proofs makes it possible to derive from the relevant inductive hypothesis that

$$\mathbf{Cafrat} \rightarrow (\Box_i \bigwedge_{x \prec y} BI_i^y \wedge \Box_i \bigwedge_{x \prec y} BI_j^y).$$

This can be used to show that we have

$$\mathbf{Cafrat} \rightarrow \neg\Box_i \neg BI_i^x, \tag{3.1}$$

from which, first

$$\mathbf{Cafrat} \rightarrow \neg\Box_i \neg\Box_i BI_i^x$$

by means of the T- and the KnStrat-axiom, and second,

$$\mathbf{Cafrat} \rightarrow BI_i^x,$$

by means of the 5-axiom, classical negation and the T-axiom.

The idea underlying a proof of 3.1 is to derive a contradiction from **Cafrat** $\wedge$ $\Box_i \neg BI_i^x$, and it is here that the rather manipulated inductive hypothesis is used. The contradiction becomes apparent as soon as it is observed that on the basis of this assumption, one would show

$$\mathbf{Cafrat} \rightarrow (\mathbf{afrat} \wedge \Box_i \bigwedge_{x \prec y} BI_i^y \wedge \Box_i \bigwedge_{x \prec y} BI_j^y \wedge \Box_i \neg BI_i^x),$$

which could be specialised to get a statement expressing that from **Cafrat** it follows that for some strategy k player i knows, first, that she plays k in the subgame generated by x, second, that k is not the inductive strategy in that subgame, and, third, that k yields strictly less than the inductive strategy in that subgame. This means that, first, knowledge of one's strategy is needed, second, some technical axioms about equivalence of strategies (taking care of subgames), and third, the fact that a generic game's utility function is an injection is essential for obtaining strict inequality. Moreover, a result about inductive strategies is needed to the effect that in generic games the inductive strategy is strictly better in some subgame given that opponents play inductively in that subgame. This is a statement with the same structural function as Lemma 3.1 (used below), connecting the BI_i^x and inequality statements with u_i^x about utility.

3.1.1.2 An Implicit, Inductive Formalisation of Rationality

This is all fine, but it is slightly cumbersome. Moreover, a proof system is used in which the T-axiom (veridicality), the 4-axiom (positive introspection) and the 5-axiom (negative introspection) figure. There is, however, no need to use such heavy machinery—endangering the general format of epistemic characterisation results in the way witnessed by that of the Nash equilibrium—once we adopt an inductive and implicit axiomatisation of rationality by means of the following three axioms.

NRat$_{bas}$ $\quad \mathbf{nrat}_i^x \rightarrow \mathrm{nsd}_i^x(A_i, A_j)$.
NRat$_{ind}$ $\quad (\mathbf{nrat}_i^x \wedge \Box_i X_i \wedge \Box_i X_j) \rightarrow \mathrm{nsd}_i^x(X_i, X_j)$.
FRat $\quad \mathbf{frat}_i \leftrightarrow \bigwedge_{\rho \preceq x} \mathbf{nrat}_i^x$.

These axioms need some explanation. First, a preliminary remark about applying on-path rationality to subgames is appropriate. On the one-shot interpretation, objects of choice (full strategies) are always functions mapping all decision nodes of a player to actions, and consequently beliefs are about the full strategies opponents choose. It makes perfect sense, though, to speak about the on-path rationality of a strategy in any subgame, for one can consider the restriction to the subgame of a strategy to evaluate its rationality as a course of action in the subgame and in light of the restrictions to the subgame of the strategies one expects one's opponents to play. This idea is captured in the first two axioms.

What is rational often depends on one's beliefs, but not always; and as I have argued, this forms the basis of the implicit and inductive formalisation of rationality. The first-axiom captures the base case without beliefs. It states that player i, if on-path rational in the subgame Γ_x generated by x, never chooses a strategy which prescribes bad actions independently of what her opponents play. If player i is on-path rational in Γ_x, she does not choose any strategy of which the restriction to Γ_x coincides with a strategy strictly dominated in the normal form of Γ_x.

The second axiom states that, if she is on-path rational in Γ_x, player i never plays a strategy that is strictly dominated in the normal form of Γ_x from which those strategies (both of her opponents as well as herself) have been removed that she believes will not be chosen. The beliefs are represented by sets X_i and X_j of strategies. Finally, the third axiom states that player i is off-path rational in the entire game if she is on-path rational in all of its subgames. The proof system ${}_{\Gamma}\mathbf{K_{EC}frat}$ used consists of axioms Prop, Dual, K, E, C, the proof rules modus ponens, necessitation and induction, all axioms for one-shot game-playing situations, plus the three above rationality axioms. A formalisation of the assumptions can be found in Table 3.1.

Table 3.1

Assumptions	
preferences	$\bigwedge_{i,k,l} \mathbf{u}_i(k,l) = \mathbf{r}_{i,k,l}$
principles	$\bigwedge_i \mathbf{frat}_i$
beliefs	
preferences	$\mathbf{C} \bigwedge_{i,k,l} \mathbf{u}_i(k,l) = \mathbf{r}_{i,k,l}$
principles	$\mathbf{C} \bigwedge_i \mathbf{frat}_i$
performed action	–
Solution Concept	
player i	BI_i
Proof System	${}_{\Gamma}\mathbf{K_{EC}frat}$

To prove the theorem we need two lemmas. To collect all propositional formulae for backward induction strategies in any subgame Γ_x, first take all strategies that prescribe backward induction actions in all real subgames of Γ_x to obtain the set $\bigcap_{x \prec y} BI_i^y$. Some of the strategies in this set, however, do not prescribe the backward induction action at x, and therefore attention has to be restricted to those elements for which there is no strictly better alternative given that all players take backward induction actions at decision nodes $y \succ x$. This is establishes the first lemma.

Lemma 3.1 $BI_i^x = \mathrm{nsd}_i^x(\bigcap_{x \prec y} BI_1^y, \ldots, \bigcap_{x \prec y} BI_N^y) \cap \bigcap_{x \prec y} BI_i^y$.

In the formalism proposed, the formula $\bigwedge_{x \prec y} \bigvee BI_i^y$ states that at any $y \succ x$ player i plays according to backward induction (if y is a decision node of hers). The intersection $\bigcap_{x \prec y} BI_i^y$ not being empty (it contains all strategies available to i in Γ that prescribe backward induction actions in Γ_x), it is simple to observe that $\vdash_{{}_{\Gamma}\mathbf{K_C frat}} \bigwedge_{x \prec y} \bigvee BI_i^y \rightarrow \bigvee \bigcap_{x \prec y} BI_i^y$. With the convention to omit disjunction symbols in front of sets of propositional formulae, this establishes the second lemma.

Lemma 3.2 $\vdash_{\Gamma\mathbf{K_C frat}} \bigwedge_{x\prec y} BI_i^y \rightarrow \bigcap_{x\prec y} BI_i^y$.

I can now turn to a proof of Theorem 3.1 in the alternative formalism. In fact, the theorem can now be phrased in terms of true belief rather than knowledge.

Proof I prove the result for $N = 2$ with players i and $j \neq i$. For more than two players one only needs to add the relevant conjuncts and to expand nsd_i^x to a function taking three or more arguments. Clearly $\mathbf{frat}_i \rightarrow \mathbf{nrat}_i^x$ by axiom FRat, and the case of a decision node x with depth $d(x) = 1$ reduces to axiom NRat_{bas} with an application of Lemma 3.1. Let $d(x) > 1$. The inductive hypothesis gives for every $y \succ x$

$$(\mathbf{Cfrat} \wedge \mathbf{frat}) \rightarrow BI_i^y.$$

Because I consider finite games, the proofs for all $y \succ x$ and both players i and j can be aggregated into

$$(\mathbf{Cfrat} \wedge \mathbf{frat}) \rightarrow (\bigwedge_{x\prec y} BI_i^y \wedge \bigwedge_{x\prec y} BI_j^y),$$

and, applying Lemma 3.2, into

$$(\mathbf{Cfrat} \wedge \mathbf{frat}) \rightarrow (\bigcap_{x\prec y} BI_i^y \wedge \bigcap_{x\prec y} BI_j^y).$$

An application of the necessitation rule for i and the K-axiom, together with some propositional reasoning, yields

$$\mathbf{Cfrat} \rightarrow (\Box_i \bigcap_{x\prec y} BI_i^y \wedge \Box_i \bigcap_{x\prec y} BI_j^y).$$

Since $\mathbf{frat}_i \rightarrow \mathbf{nrat}_i^x$, we find

$$(\mathbf{Cfrat} \wedge \mathbf{frat}) \rightarrow (\mathbf{nrat}_i^x \wedge \Box_i \bigcap_{x\prec y} BI_i^y \wedge \Box_i \bigcap_{x\prec y} BI_j^y).$$

Applying the NRat_{ind}-axiom gives

$$(\mathbf{Cfrat} \wedge \mathbf{frat}) \rightarrow \text{nsd}_i^x(\bigcap_{x\prec y} BI_i^y, \bigcap_{x\prec y} BI_j^y).$$

Invoking the inductive hypothesis again, and applying Lemma 3.2, the consequent of this formula can be made somewhat more precise in

$$(\mathbf{Cfrat} \wedge \mathbf{frat}) \rightarrow \text{nsd}_i^x(\bigcap_{x\prec y} BI_i^y, \bigcap_{x\prec y} BI_j^y) \cap \bigcap_{x\prec y} BI_i^y,$$

which is

$$(\mathbf{Cfrat} \wedge \mathbf{frat}) \rightarrow BI_i^x$$

by Lemma 3.1.

3.1.2 Discussion

I have presented two versions of the one-shot interpretation of extensive game-play: one with on-path rationality and one with off-path rationality. Aumann was seen to adopt the latter version. I do not believe, however, that the latter version is conceptually consistent; off-path rationality is strictly incompatible with the true spirit of the one-shot interpretation. The reason is that there is no sensible rationale to take care of what would happen at unreached, off-path nodes in a situation in which the objects of choice are strategies from the normal form of an extensive game. In a one-shot situation, it just does not make sense to talk about nodes being reached or not. The game-playing situation is a strategic predicament in which the players choose a strategy that fixes a complete plan of action for the entire game. Temporal deliberation is senseless, as is thinking about players having beliefs at various points in a temporally extended sequence of decision moments. No nodes are reached or unreached. There is only one decision moment and the outcome of the game is determined on the basis of the strategies the players choose at that precise decision moment.

Does the fact that the one-shot interpretation leaves no room for rationality notions transcending the normal form entail that the epistemic characterisation result of backward induction fails to be significant, or that backward induction cannot be epistemically characterised in a one-shot interpretation? I answer the first question in the affirmative. There is no sense to any epistemic characterisation that presupposes the one-shot interpretation together with a form of rationality that goes beyond on-path rationality by using the specific structural properties of extensive games. The second question, however, need not be answered in the affirmative. It is not difficult to see that once you rephrase the NRat-axioms in terms of weak rather than strict dominance, backward induction can be characterised on the basis of on-path rationality at the root of the game. Given an extensive game with perfect information Γ, let proof system ${}_{\Gamma}\mathbf{K_C Nrat'}$ consist of the following axioms: Prop, Dual, K, C, the proof rules modus ponens, necessitation and induction, all axioms for one-shot game-playing situations for Γ, plus the following two rationality axioms.

NRat'_{bas} $\quad \mathbf{Nrat'}^x_i \rightarrow \text{nwd}^x_i(A_i, A_j)$.
NRat'_{ind} $\quad (\mathbf{Nrat'}^x_i \wedge \Box_i X_i \wedge \Box_i X_j) \rightarrow \text{nwd}^x_i(X_i, X_j)$.

The following theorem captures the relation between backward induction and common true belief about rationality in terms of weak dominance.

Theorem 3.2 *Let Γ be a finite generic N-person extensive game with perfect information. Then*

$$\vdash_{\Gamma \mathbf{K_C Nrat'}} (\mathbf{CNrat'}^{\rho} \wedge \mathbf{Nrat'}^{\rho}) \rightarrow \bigwedge_i BI^{\rho}_i .$$

To prove this theorem, first observe that on the level of the normal form of the extensive game, the relevant solution concept is iterated weak dominance. Although the actual outcome of a process of iterative elimination of weakly dominated strategies depends on the exact definition of the elimination algorithm, a lemma due to Hervé Moulin shows this to be irrelevant for current purposes.[8]

Lemma 3.3 *Let Γ be the a finite generic N-person extensive game with perfect information, and let* $\mathrm{nf}(\Gamma)$ *be its normal form. Then*

1. *Any* natural *algorithm for iterated weak dominance yields precisely one strategy profile in* $\mathrm{nf}(\Gamma)$.
2. *All of these algorithms yield the same strategy profile.*
3. *The strategies from this profile correspond to the backward induction strategies of* Γ.

The proof of Theorem 3.2 is a direct analogue of the proof of Theorem 3.1. This precipitates us, of course, into the problem encountered earlier concerning the dubious status of the solution concept of iterated weak dominance. I will not reiterate this point, but instead proceed to the many-moment interpretation.

3.2 The Many-Moment Interpretation

3.2.1 The Inconsistency Result

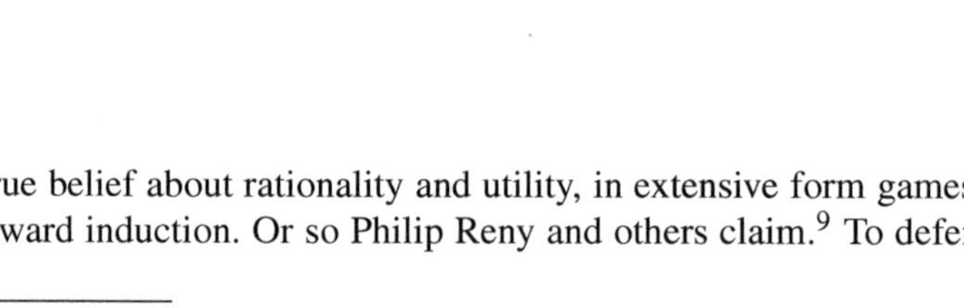

Fig. 3.1

Common true belief about rationality and utility, in extensive form games, does not yield backward induction. Or so Philip Reny and others claim.[9] To defend this

[8] *Game Theory for the Social Sciences* (New York: New York University Press, 1986).

[9] Philip Reny, art. cit. Kaushik Basu, 'Strategic Irrationality in Extensive Games', *Mathematical Social Sciences*, 15 (1988), 247–260 investigates temporality aspects. Elchanan Ben-Porath, 'Rationality, Nash Equilibrium and Backwards Induction in Perfect-Information Games', *Review of Economic Studies*, 64 (1997), 23–46 develops an account in terms of common certainty. Christina Bicchieri, 'Common Knowledge and Backward Induction: A Solution to the Paradox', in M. Vardi (ed.), *Proceedings of the Second Conference on Theoretical Aspects of Reasoning about Knowledge: March 7–9, 1988, Pacific Grove, California* (Los Altos, Calif.: Morgan Kaufmann,

claim, Reny proves an inconsistency result in the context of the many-moment interpretation of extensive game-play where, in contrast to the one-shot interpretation, games are seen as models of a temporal succession of many decision moments. Referring to the game shown in Figure 3.1, for instance, Reny writes that

> I claim that if player one does not take the dollar and end the game in the first round [does not play D_1], but instead leaves it so that player 2 must decide whether or not to take the two dollars [whether or not to play d_1], then it is no longer possible for rationality to be common knowledge. (i.e. At [*sic*] player two's information set, it is not possible for rationality to be common knowledge).[10]

Such reasoning makes no sense in the one-shot interpretation, according to which no decision nodes are reached at all. On the contrary, strategies are chosen which may or may not induce a path through the game tree to reach some decision node. But it does not make sense to talk about the beliefs of the players at those decision nodes. The players have beliefs at the moment they choose their strategy, but the game stops after that. Reny, by contrast, considers beliefs of players at some decision moment—typical of the many-moment interpretation. Such beliefs describe the expectations of a player about what actions will be taken at decision nodes in the subgame generated by the current decision node.

In principle, as I have argued, the many-moment interpretation leaves two possibilities open. The beliefs at some decision moment can, first, be viewed as dependent on what happened before, and be sensitive to the history of the decision moment in the sense, for instance, that a player could decide to believe her opponent to be irrational if the current decision node can only be reached by the irrational play of her opponent. Second, the beliefs can be viewed as completely insensitive to history. It will be seen that Reny's inconsistency result presupposes a history-sensitive view of belief formation.

Before presenting a formalisation of Reny's inconsistency theorem criticising the epistemic characterisation of backward induction in terms of common true belief about rationality and utility, I should spell out the purpose of the present section, and set down some notation. In the preceding section I developed a logical framework for the one-shot interpretation of extensive games and applied it to Aumann's result on common knowledge of rationality and backward induction. Likewise, I here de-

---

1988), 381–393, and ead., 'Self-Refuting Theories of Strategic Interaction: A Paradox of Common Knowledge', *Erkenntnis*, 30 (1989), 69–85 develop the view that players can be seen as possessing theories reflecting the epistemic set-up of game-playing. Kenneth Binmore, 'Rationality and Backward Induction', *Journal of Economic Methodology*, 4 (1997), 23–41 zooms in on counterfactual reasoning. Thorsten Clausing, 'Doxastic Conditions for Backward Induction', *Theory and Decision*, 54 (2003), 315–336 sets up a truly doxastic system related to ours. Robert Stalnaker, 'Belief Revision in Games: Forward and Backward Induction', *Mathematical Social Sciences*, 36 (1998), 31–56 studies these issues from the perspective of game models and belief revision theory. More philosophical essays on backward induction include Philip Pettit and Robert Sugden, 'The Backward Induction Paradox', *The Journal of Philosophy*, 86 (1989), 169–182, Jordan Howard Sobel, 'Backward-Induction Arguments: A Paradox Regained', *Philosophy of Science*, 60 (1993), 114–133, and Roy Sorensen, 'Paradoxes of Rationality', in A. Mele (ed.), *The Handbook of Rationality* (Oxford: Oxford University Press, 2004).

[10] Reny, 'Common Knowledge and Games with Perfect Information', 364–365.

velop a logical framework for the many-moment interpretation. Rather than using it to characterise backward induction from the point of view of the many-moment interpretation, I turn to an inconsistency theorem which denies that common knowledge of rationality entails backward induction. It should be stressed, though, that the many-moment perspective has also been adopted to defend the implication of backward induction by common knowledge of rationality. While such forms of defence characterise backward induction in subclasses of extensive games only, they do not make the dubious assumptions that underlie the inconsistency result that I ultimately reject in this section.[11]

Recall that I use super-scripted beliefs to indicate beliefs at decision moments at which the respective decision is reached. Then first define, on the basis of beliefs $\mathbf{P}_i^x(\mathbf{i}_k^x) = \mathbf{p}_k$ and $\mathbf{P}_i^x(\mathbf{j}_l^x) = \mathbf{p}_l$ an auxiliary notion of the expected utility conditional on reaching some immediate successor $y \succ x$ as

$$\mathrm{EU}_i(y, \mathbf{P}_i^x) = \sum_{k,l} \mathbf{p}_k \mathbf{p}_l \mathbf{u}_i^y(k,l).$$

Then define $\mathrm{EU}_i(k,x,\mathbf{P}_i^x)$, the intended interpretation being the expected utility of playing according to the kth strategy at the decision moment at which node x is reached, as

$$\mathrm{EU}_i(k,x,\mathbf{P}_i^x) = \mathrm{EU}_i(y,\mathbf{P}_i^x)$$

for the immediate successor y that is reached when at x player i plays according to his kth strategy.

3.2.1.1 An Explicit Formalisation of Rationality

To formalise rationality, the principle of expected utility maximisation can be relativised to subgames thus.

RRat $\mathbf{rrat}_i^x \leftrightarrow ((\Box_i^x \bigwedge_{k,l} \mathbf{u}_i^x(k,l) = \mathbf{r}_{i,k,l} \wedge \bigwedge_k \mathbf{P}_i^x(\mathbf{i}_k^x) = \mathbf{p}_k \wedge \bigwedge_l \mathbf{P}_i^x(\mathbf{j}_l^x) = \mathbf{p}_l \wedge \mathbf{i}_m(x)) \rightarrow \bigwedge_k \mathrm{EU}_i(m,x,\mathbf{P}_i^x) \geq \mathrm{EU}_i(k,x,\mathbf{P}_i^x))$.

The antecedent of the right hand side contains a condition on the beliefs about the utility structure, and on probabilistic beliefs about what player i herself and her opponent j will play. In the consequent it is stated that i will maximise her expected utility given her beliefs.

[11] Many-moment advocates of the implication of backward induction by common knowledge of rationality (or even by a weakening of these assumptions) include Robert Aumann, 'On the Centipede Game', *Games and Economic Behavior*, 23 (1998), 97–105, Wlodek Rabinowicz, 'Grappling with the Centipede: Defence of Backward Induction for BI-terminating Games', *Economics and Philosophy*, 14 (1999), 95–126, John Broome and Wlodek Rabinowicz, 'Backwards Induction in the Centipede Game', *Analysis*, 59 (1999), 237–242, and Magnus Jiborn and Wlodek Rabinowicz, 'Backward Induction without Full Trust in Rationality', in W. Rabinowicz (ed.), *Value and Choice: Some Common Themes in Decision Theory and Moral Philosophy: Volume 2* (Lund: Lund Philosophy Reports, 2001), 101–120.

We need additional axioms, however, to fix the belief formation policies of the players. First, players do not revise their beliefs during game-play as long as this does not lead to inconsistency.

StratPers $\quad \bigwedge_i \bigwedge_j \mathbf{P}_i^x(\mathbf{j}_k^z) = \mathbf{P}_i^y(\mathbf{j}_k^z)$,

for $x \preceq y \preceq z$. This persistence axiom states that if $x \preceq y \preceq z$, then the beliefs that player i has at x about the action of her opponent or herself at z will be the same at y. Of course, if game-play has passed z and the beliefs have been contradicted, then i will have different beliefs. But as long as z has not been reached the beliefs remain constant.

While this axiom concerns beliefs about strategies, we need another axiom that involves beliefs about rationality. It states that a player never gives up her beliefs about someone's rationality as long as that person has not moved; in more technical language, the axiom states that if i believes at x that j is rational at some future node y, then i will not change that belief as long as j has not moved.

RatPers $\quad \bigwedge_i \bigwedge_j (\Box_i^x \mathbf{rrat}_j^y \leftrightarrow \Box_i^x \mathbf{rrat}_j^z)$,

where $x \preceq y \prec z$, $\iota(z) = j$, and no u with $\iota(u) = j$ exists such that $y \prec u \prec z$. It is left to the reader to verify that a striking consequence of this is that either i believes j to be rational everywhere, or nowhere.

Table 3.2

Assumptions	
preferences	$\bigwedge_{i,k,l} \mathbf{u}_i(k,l) = \mathbf{r}_{i,k,l}$
principles	–
beliefs	
preferences	$\mathbf{C} \bigwedge_{i,k,l} \mathbf{u}_i(k,l) = \mathbf{r}_{i,k,l}$
principles	–
performed action	–
Inconsistency	
node x	$\neg \mathbf{C}^x \mathbf{rrat}^x$
Proof System	$_\Gamma\mathbf{KD_{EC}rrat}$

To prove the inconsistency result, the proof system $_\Gamma\mathbf{KD_{EC}Prrat}$ with Prop, Dual, K, D, E, C, the linear (in)equality axioms, the Kolmogorov axioms, the interrelation axioms, the proof rules modus ponens, necessitation and induction, all axioms for many-moment game-playing situations, the rationality axiom and the two persistence axioms are used. The formalisation of the assumptions can be found in Table 3.2.

Theorem 3.3 (Reny, 1988) *There is an extensive form game with perfect information such that for all game-playing situations that consist of at least two decision* *moments there cannot be common true belief, at the second decision moment, among the players that they are rational.*

Reny's original proof involves an argument to the effect that no game-playing situation of the game shown in Figure 3.1 can have common belief about rationality at its second moment, because every second decision moment would be a moment at which at x_1 player 2 has to move. In this original game, there is no way for both players to play on and gain (only one will gain from playing on) and hence the suggestion might arise that the inconsistency result is not too surprising after all. However, the result can be proven in a case where both players would gain from playing on, too, and to underline this I prove the result using the game shown in Figure 3.2 rather than Reny's original game shown in Figure 3.1. In fact, Reny provides an argument which reveals that the class of games for which inconsistency results can be proven is fairly large.[12]

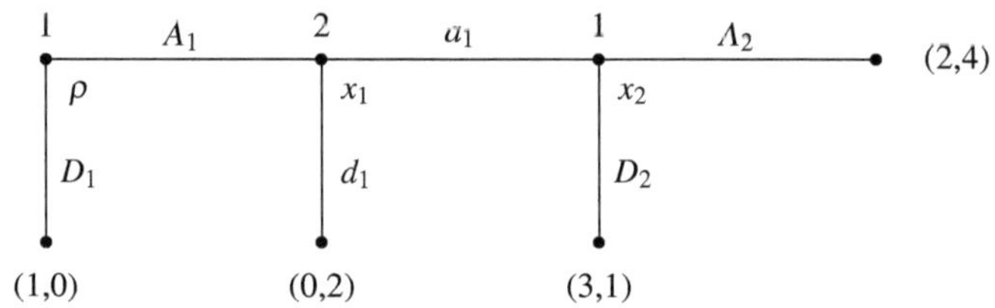

Fig. 3.2

Proof I make use of the two interrelation axioms Cons and KnProb, because all relevant beliefs in this proof all have probability one, and I use obvious notation to refer to strategies. I first prove

$$\mathbf{C}^{x_1}\mathbf{rrat}^{x_1} \to \Box_2^{x_1}\Box_1^{p} d_1. \tag{3.2}$$

To do that, it suffices to show

$$\mathbf{C}^{x_1}\mathbf{rrat}^{x_1} \to \Box_1^{p} d_1, \tag{3.3}$$

because a simple argument using the rule of necessitation for $\Box_2^{x_1}$ concludes the proof. Because of the StratPers-axiom, however, to prove 3.3, it suffices to show

$$\mathbf{C}^{x_1}\mathbf{rrat}^{x_1} \to \Box_1^{x_1} d_1, \tag{3.4}$$

and to show 3.4, in turn, it is shown that

---

[12] 'Common Belief and the Theory of Games with Perfect Information', *Journal of Economic Theory*, 59 (1993), 269.

$$\mathbf{C}^{x_1}\mathbf{rrat}^{x_1} \to (\Box_1^{x_1}\mathbf{rrat}_2^{x_1} \wedge \Box_1^{x_1}\Box_2^{x_1}D_2), \tag{3.5}$$

and then apply necessitation for $\Box_1^{x_1}$ to an instance of the rationality axiom to get

$$(\Box_1^{x_1}\mathbf{rrat}_2^{x_1} \wedge \Box_1^{x_1}\Box_2^{x_1}D_2) \to \Box_1^{x_1}d_1.$$

The remainder of the proof of 3.2 is devoted to showing 3.5. Clearly we have

$$\mathbf{C}^{x_1}\mathbf{rrat}^{x_1} \to \Box_1^{x_1}\mathbf{rrat}_2^{x_1}.$$

To prove

$$\mathbf{C}^{x_1}\mathbf{rrat}^{x_1} \to \Box_1^{x_1}\Box_2^{x_1}D_2,$$

observe that with the RatPers-axiom for $\Box_2^{x_1}$ and necessitation for $\Box_1^{x_1}$ it can be shown that

$$\Box_1^{x_1}\Box_2^{x_1}\mathbf{rrat}_1^{x_1} \to \Box_1^{x_1}\Box_2^{x_1}\mathbf{rrat}_1^{x_2},$$

because x_2 is a successor of x_1 at which 1 moves for which in addition no y with $x_1 \succ y \succ x_2$ exists at which it is 2's turn. Hence

$$\mathbf{C}^{x_1}\mathbf{rrat}^{x_1} \to \Box_1^{x_1}\Box_2^{x_1}\mathbf{rrat}_1^{x_2}.$$

Applying the rationality axioms concludes the proof of 3.2.

Observe now that it is an easy consequence of the RatPers-axiom that

$$\mathbf{C}^{x_1}\mathbf{rrat}^{x_1} \to \Box_2^{x_1}\mathbf{rrat}_1^{\rho}, \tag{3.6}$$

and that

$$(\Box_2^{x_1}\Box_1^{\rho}d_1 \wedge \Box_2^{x_1}\mathbf{rrat}_1^{\rho}) \to \Box_2^{x_1}\neg A_1. \tag{3.7}$$

follows directly from the rationality axiom plus an appropriate application of the rule of necessitation that. All this is used to prove

$$\mathbf{C}^{x_1}\mathbf{rrat}^{x_1} \to \bot. \tag{3.8}$$

The KnWhere-axiom gives $\Box_2^{x_1}A_1$. Hence it suffices to show that

$$\mathbf{C}^{x_1}\mathbf{rrat}^{x_1} \to \neg\Box_2^{x_1}A_1.$$

Combining 3.2 and 3.6 gives

$$\mathbf{C}^{x_1}\mathbf{rrat}^{x_1} \to (\Box_2^{x_1}\Box_1^{\rho}d_1 \wedge \Box_2^{x_1}\mathbf{rrat}_1^{\rho})$$

to which application of 3.7 gives

$$\mathbf{C}^{x_1}\mathbf{rrat}^{x_1} \to \Box_2^{x_1}\neg A_1.$$

An application of the D-axiom finishes the proof.

3.2.2 Discussion

Given the game-theoretic view of rationality as expected utility maximisation, the question to ask is not so much whether the RRat-axiom is plausible, but whether the belief persistence principles embodied in StratPers and RatPers are plausible. Let me distinguish the plausibility of the principles in general, and the plausibility of the specific instances in the proof of Theorem 3.3.

Let me start with the general plausibility of the StratPers-axiom:

$$\bigwedge_i \bigwedge_j \mathbf{P}_i^x(\mathbf{j}_k^z) = \mathbf{P}_i^y(\mathbf{j}_k^z),$$

for $x \preceq y \preceq z$. A possible argument in favour of this principle runs as follows. If at x player i believes that at some $z \succeq x$ her opponent j chooses action a, say, then there is no need for i to revise her beliefs at some intermediate y (satisfying $x \preceq y \preceq z$, that is) as long as i has not received any contradictory information on the way from x to y. But information contradicting that j chooses a can only be information that j chooses, at z, an action different from a. Player i has not received such information at the intermediate y, and consequently she will not need to revise her beliefs at y. Arguably, this yields a defence of StratPers.

Yet this argument overlooks subtle ways of obtaining pertinent information. A reason for player i's belief that j chooses a at z may be her belief that at z player j chooses rationally. If, however, on the path from x to y, player i has seen j choosing irrationally, this reason is probably no longer available. Player i might revise her beliefs in such a way that at z player j plays irrationally, too—not choosing a. She need not change her beliefs, but she may change them, and that is sufficient to make StratPers problematic.

This is a general problem with the StratPers-axiom. It completely ignores the fact that the reasons players have for particular beliefs may change over time, and that consequently they have to reconsider (or revise) their beliefs, even if they are not directly contradicted by observed facts.

Similar arguments work against the general plausibility of the RatPers-axiom:

$$\bigwedge_i \bigwedge_j (\Box_i^x \mathbf{rrat}_j^y \leftrightarrow \Box_i^x \mathbf{rrat}_j^z),$$

where $x \preceq y \prec z$, $\iota(z) = j$, and no u with $\iota(u) = j$ exists such that $y \prec u \prec z$. Imagine that only irrational play on the part of j may get her from y to z. Although i believes, at x, that j will not take that irrational route, it is still questionable whether i should maintain that even though j plays irrationally, she will return to playing rationally at node z.

Now it may be that the use of the belief persistence axioms in the proof of Theorem 3.3 is harmless. StratPers is used to prove

$$\Box_1^{x_1} d_1 \rightarrow \Box_1^{\rho} d_1$$

in the proof of 3.2. The general difficulty that displays is clearly revealed. The reasons for the belief $\Box_1^{x_1} d_1$ are the beliefs $\Box_1^{x_1}\mathbf{rrat}_2^{x_1} \wedge \Box_1^{x_1}\Box_2^{x_1} D_2$. This is so because $(\Box_1^{x_1}\mathbf{rrat}_2^{x_1} \wedge \Box_1^{x_1}\Box_2^{x_1} D_2) \rightarrow \Box_1^{x_1} d_1$ is obtained by necessitation on an instance of the RRat-axiom. These reasons, perhaps available at x_1, may not be available at ρ, though. That is, it may be doubted whether $\Box_1^{\rho}\mathbf{rrat}_2^{x_1} \wedge \Box_1^{\rho}\Box_2^{x_1} D_2$. One way to substantiate doubt would concern the second conjunct. As inspection of the proof of 3.2 shows, the reasons for $\Box_1^{x_1}\Box_2^{x_1} D_2$ involve, among other things, player 1's beliefs about 2's beliefs about the rationality of player 1 at x_2, or $\Box_1^{x_1}\Box_2^{x_1}\mathbf{rrat}_1^{x_2}$. The question whether these may figure as reasons for the beliefs at the root of the game (reasons for $\Box_1^{\rho} d_1$) then boils down to the question whether these reasons were already available at the root of the game; that is, whether $\Box_1^{\rho}\Box_2^{x_1}\mathbf{rrat}_1^{x_2}$ follows from $\Box_1^{x_1}\Box_2^{x_1}\mathbf{rrat}_1^{x_2}$.

It may come as an anticlimax that there do not seem to be any serious problems here. It is about player 1 imagining (at the root and at x_1) what player 2 will or does believe at x_1 about player 1 at x_2. But player 1 will not have obtained any new information about player 2's beliefs (at x_1) while going from the root to x_1. At the root, player 1 imagines player 2's beliefs at x_1, and at x_1 player 1 again imagines player 2's beliefs at x_1. There is no difference between these cases. There would have been a difference had the statement compared player 2's beliefs at the root with his beliefs at x_1. But that is not the issue here. Consequently, StratPers causes no harm to the plausibility of the assumptions of Theorem 3.3.

To turn to the RatPers-axiom, it is first used to arrive at

$$\Box_1^{x_1}\Box_2^{x_1}\mathbf{rrat}_1^{x_1} \rightarrow \Box_1^{x_1}\Box_2^{x_1}\mathbf{rrat}_1^{x_2}$$

in the proof of 3.5. You may find this problematic as it involves the rationality of player 1 at a decision moment where she need not choose any action. But apart from that there do not seem to be reasons to doubt this line of reasoning. Player 1 does not move at x_1, so player 2, if he believes that 1 is rational at the decision moment corresponding to x_1, has no reason to say that 1 would not be rational at the possible succeeding decision moment. Player 1 believes all this, and consequently the RatPers-axiom is unproblematic here.

Yet it is also used to prove 3.6,

$$\mathbf{C}^{x_1}\mathbf{rrat}^{x_1} \rightarrow \Box_2^{x_1}\mathbf{rrat}_1^{\rho},$$

and here I can point to something dubious: a belief revision policy, forced upon player 2, that is excessively rigid. It excludes, for instance, sensible dealings with a situation of the following kind. Player 2 has actually arrived at x_1, so player 1 has moved across. While player 2 considers this to be irrational, he also believes it to be an accident or a mistake. At x_2, that is, player 2 believes that player 1 was irrational at the first decision moment, but he also believes at x_2 that player 1 is rational at the second decision moment (and perhaps even the third). This kind of subtle belief revision policy is excluded by RatPers. Either a player is believed to be rational everywhere or irrational everywhere.

In summary, the proof of Theorem 3.3 boils down to showing that there is a contradiction between having arrived at x_1 and there being common true belief about rationality at x_1. Such a contradiction can only be shown if, from the fact that there is common belief about rationality at x_1, it can be derived that one cannot be at x_1, more specifically, that one cannot be at x_1 because it can only be reached irrationally. That can only be demonstrated successfully if, from common belief about rationality at x_1, something follows about the beliefs and rationality at ρ.

But there is nothing in the concept of common belief at some decision moment that obliges me to interpret it in such a temporally extended way, and to adopt corresponding belief revision policies. In other words, there is nothing to disallow a game-playing situation in which at the first decision moment there is no common belief about rationality, while there is in the second. The RatPers axiom (together with the StratPers-axiom) exclude that possibility. This means that they are too strict. That being the case, the inconsistency result only works under very heavy, if not implausible assumptions, and shares this fate with the epistemic characterisation result it set out to criticise.

Part II
Epistemology

Chapter 4
Applications of Game Theory

The aim of this book is to contribute to the philosophy of the theory of games by offering internal as well as external investigations of game theory. The internal investigations were reported in the first three chapters, especially in the discussions in Chapters 2 and 3, and in places made use of heavily technical, logical apparatus. Among the outcomes were that the Nash equilibrium implausibly presupposes veridicality of beliefs, that the Dekel–Fudenberg procedure is not necessarily the only consequence of common true belief about payoff-uncertainty, and that both Aumann's and Reny's analyses of backward induction make conceptually inconsistent assumptions.

It is now time to adopt an external point of view. The present chapter further develops the insight into the logical form of game theory, as it is used descriptively and normatively, and points to a number of serious problems here. It is argued that game-theoretic explanations always implicitly refer to the epistemic set-up revealed by the epistemic characterisation results, but that on that account a rather narrow-minded process of belief formation is ascribed to the players. It is argued, too, that as a normative theory game theory gives bad advice, is reducible to decision theory, or is plain nonsensical. This does not mean that non-cooperative game theory is entirely useless, however.

While taking an external look at game theory involves criticism, I am not concerned here with the merits and flaws of concrete applications of game theory in the social sciences. Psychologists and sociologists have raised doubts about the adequacy of decision-theoretic modelling, and have attempted to develop models that are somewhat more pertinent.[1] The criticism in the last two chapters of the book does not actually depend on particular forms of evidence against particular applications of decision and game theory. Rather, it is a kind of meta-argument pointing out

[1] Recent publications include Margaret Archer and Jonathan Tritter (eds.), *Rational Choice Theory: Resisting Colonisation* (London: Routledge, 2000), Ian Shapiro, *The Flight from Reality in the Human Sciences* (Princeton: Princeton University Press, 2005), and Michael Taylor, *Rationality and the Ideology of Disconnectedness* (Cambridge: Cambridge University Press, 2006). Paul Moser (ed.), *Rationality in Action: Contemporary Approaches* (Cambridge: Cambridge University Press, 1996) collects a number of classic essays.

B. de Bruin, *Explaining Games*, Synthese Library 346,
DOI 10.1007/978-1-4020-9906-9_4, © Springer Science+Business Media B.V. 2010

the inconsistencies and counterproductive research strategies employed by various researchers.

4.1 Logical Form

4.1.1 Rationality

It is sometimes said that explanations in decision and game theory are empty because the underlying notion of rationality is empty; that is, if an agent chooses to perform one action rather than another, she believes it (at least at the time she chooses it) to be the best choice given her beliefs. That she performs it shows that it maximises her expected utility. Others make a very similar point in a more technical way and state that agents can always be described 'as if' they maximise their expected utility, adducing various mathematical results about *revealed preferences* to support their views.[2] But whether the naive or the sophisticated version of the emptiness claim is adopted, according to this view you just have to say that agents' actions maximise their expected utility because they perform them, rather than holding to the converse view that agents perform actions because they maximise expected utility.

Instrumentalism may be widespread in economic theory, but I think there is much to recommend taking a different point of view.[3] An instrumentalist cannot sensibly distinguish between the merits of two models if they give the same prediction on the basis of the same data. If models are *input–output equivalent*, for the instrumentalist they are the same. Several models of market entrance, for instance, are input–output equivalent, but they differ internally, some postulating lack of common knowledge

---

[2] I thank Frank Hindriks for a question about this point. The presentation owes much to Daniel Hausman, 'Revealed Preference, Belief, and Game Theory', *Economics and Philosophy*, 16 (2000), 99–115. Revealed preference theory was developed by Paul Samuelson, 'A Note on the Pure Theory of Consumers' Behaviour', *Economica*, 5 (1938), 61–71. Defining preferences in terms of dispositions to choose, the idea is that in order to find out what an agent's preference ordering is, she is confronted with a number of choice problems. The axioms of revealed preference then ensure that under certain conditions the underlying preference ordering of the agent can be derived. The term *revealed preference theory* is also used to refer to the idea that the only access the theorist has to an agent's preferences and beliefs is observation of her actual action. Even if this were plausible (which I do not think, given what historians learn from reading diaries of historical figures) it would not follow that one can derive someone's beliefs and desires merely from information about rationality; an action maximises expected utility relative to a whole lot of combinations of expectations and utility functions.

[3] The classic reference is Milton Friedman, 'The Methodology of Positive Economics', in ibid., *Essays in Positive Economics* (Chicago: University of Chicago Press, 1953), 3–43. In game theory, the most explicit phrasing is probably due to Robert Aumann, 'What is Game Theory Trying to Accomplish?', in K. Arrow and S. Honkapohja (eds.), *Frontiers of Economics* (Oxford: Blackwell, 1987), 28–76.

of rationality, others postulating lack of common knowledge of utility structure.[4] If the Epistemic Programme in game theory is to make any sense, however, these differences should matter. Common knowledge of rationality and common knowledge of utility are two different elements of a game-playing situation. Why bother to investigate the details of payoff-uncertainty if the same output can be obtained from uncertainty about rationality? The answer is, at least for many researchers working in the Epistemic Programme, that in numerous concrete game-playing situations there is a fact to the matter about whether someone is partly misinformed about her opponents' utility functions, or about their rationality principles. This is not a knock-down argument against instrumentalism, and perhaps such an argument is just impossible to make. Yet if game theory is to have any intellectual bite, it has to be thought of, not as making merely instrumentalist claims, but rather as putting forward *realist* statements about players, their payoffs, their rationality principles, and their beliefs. Yet, I should stress that while I consider rationality to be an essential notion—rather than an 'as if' fiction—in giving an account of human behaviour, the two main claims about game-theoretic explanations and norms defended in this chapter do not depend on this view. The claims do depend on the logical analysis below, but this analysis works equally well with a purely instrumentalist view of rationality.

In order to put the game-theoretic concept of rationality as expected utility maximisation into perspective, I will first look at a broad, almost metaphysical or methodological conception of rationality in the works of Max Weber. I will then turn to a narrower, rather empirical notion that can be found in John Stuart Mill's writings, and these preliminaries allow me to locate expected utility maximisation with more precision at the end.

4.1.1.1 Max Weber and John Stuart Mill

To get a feel for what is meant here, it is instructive to look first at causality. A metaphysical reading of causality is found in the statement that all events have causes; the defence of this statement involves conceptual, metaphysical argumentation. Loosely following Karl Popper, this metaphysical proposition can be turned into a methodological principle when you formulate the rule that all good explanations of events have to specify their causes, and this rule may, in turn, be rendered into the empirical statement that causes, as causes, can truly be individuated in nature (it does not matter for now whether that statement has any plausibility).[5]

[4] Reinhard Selten, 'The Chain-Store Paradox', *Theory and Decision*, 9 (1978), 127–159 pointed out that a straightforward model of market entrance involving backward induction led to predictions that were highly off the mark. David Kreps, et al., 'Rational Cooperation in the Finitely Repeated Prisoners' Dilemma', *Journal of Economic Theory*, 27 (1982), 245–252, ibid. and Robert Wilson, 'Reputation and Imperfect Information', *Journal of Economic Theory*, 27 (1982), 253–279, and Paul Milgrom and John Roberts, 'Predation, Reputation, and Entry Deterrence', *Journal of Economic Theory*, 27 (1982), 280–312 construct alternative game-theoretic models.

[5] Karl Popper, *Logik der Forschung* (Vienna: Julius Springer, 1935). Philippe Mongin, 'Le principe de rationalité et l'unité des sciences sociales', *Revue Économique*, 53 (2002), 301–323 argues that

Causality relates events to causes, whereas rationality relates human actions to reasons. A plausible analogous metaphysical reading of rationality states that all human actions have reasons, the corresponding methodological rule requires that explanations of actions refer to the reasons of the agents, and the empirical counterpart is to the effect that human agents truly act on reasons.

What does this mean? Max Weber famously introduced four different kinds of action.[6] *Value rational* actions are guided by, for instance, religious, aesthetic or moral principles; *goal rational* actions are guided by the comparison of means and ends; *affective* actions are governed by affects and emotions, while *traditional* actions are directed by traditions. That is, there are four kinds of things that count as reasons for actions, described in terms of religious principles, goals or purposes, emotions and traditions. On the empirical level this entails a commitment to the claim that at least some actions are guided by such factors, whereas on the methodological level a requirement is put forward to explain actions in terms of Weber's four kinds of reasons.

Whether you look at the Weberian conception as a methodological call for a particular kind of explanation, or as an empirical claim about the real character of human agency, it is a rather broad one in that it allows for diversity of reasons. John Stuart Mill put forward a much narrower conception of rationality.[7] Mill suggests that in contrast with the natural sciences, the social sciences have to start from certain abstract and simplified versions of the phenomena to be explained. By '*à priori* [*sic*]' reasoning, the social scientist arrives at statements that are 'only true, as the common phrase is, *in the abstract*':

> Geometry presupposes an arbitrary definition of a line, 'that which has length but not breadth'. Just in the same manner does Political Economy presuppose an arbitrary definition of man, as a being who invariably does that by which he may obtain the greatest amount of necessaries, conveniences, and luxuries, with the smallest quantity of labour and physical self-denial with which they can be obtained in the existing state of knowledge.[8]

It is not difficult to see the conception of rationality embodied in this definition as a concrete kind of Weberian goal rationality. For Weber, a person acts according to goal rationality whenever she orients her actions 'towards goals, means, and

---

the principle of rationality Popper speaks about has to be considered as purely metaphysical, because no interesting statements can be deduced from it. On Popper's views on the principle of rationality in the social sciences, see Popper's 1963 lecture 'The Myth of the Framework', in ibid., *Models, Instruments and Truth*, ed. M. Notturno (London: Routlegde, 1998), 154–184. A critical appraisal is Boudewijn de Bruin, 'Popper's Conception of the Rationality Principle in the Social Sciences', in I. Jarvie, K. Milford and D. Miller (eds.), *Karl Popper: A Centenary Assessment: Selected Papers from Karl Popper 2002: Volume III: Science* (Aldershot: Ashgate, 2006), 207–215.

[6] Max Weber, *Wirtschaft und Gesellschaft*, ed. Marianne Weber (Tübingen: Mohr, 1921; 5th ed. 1990), 12–13.

[7] John Stuart Mill, 'On the Definition of Political Economy; and on the Method of Philosophical Investigation in that Science', *London and Westminster Review*, 26 (1836), 1–29; citations are from 1844 ed. repr. in ibid., *Collected Works of John Stuart Mill*, ed. J. Robson (Toronto: University of Toronto Press, 1967), 309–339.

[8] Ibid. 326.

side-effects, weighing rationally the means against the goals, the goals against the side-effects, and finally, too, the possible different goals against each other'.[9] Mill's 'arbitrary definition of man' is fruitfully seen as restricting not so much the aims or goals themselves as the way in which they are evaluated. Agents adjudicate between different aims solely by examining how they fare with respect to the 'amount of necessaries, conveniences, and luxuries'. Similarly, it is not the possible means that are restricted, but the ways in which they have to be compared; the less they involve 'labour and physical self-denial', the better the means are. In clear contradistinction to Weber, Mill attempts to give a full specification of the aspects of the possible aims that might be important in the process of weighing up, and it is noteworthy that he refers to what is conspicuously missing from Weber's account—the agent's epistemic state.

4.1.1.2 Decision and Game Theory

One might doubt the plausibility of the suggestion that Mill has put forward an empirical notion of rationality here, because his emphasis is clearly on the a priori, abstract, possibly metaphysical character of economics. I have represented Mill's position as empirical, though, because his notion of rationality, if taken empirically, forms a conceptually intermediate step between Weberian rationality and the rationality of decision and game theory, expected utility maximisation, to which I will now turn.

Expected utility maximisation stipulates particular kinds of means and ends, and particular ways to compare them. It applies to agents who possess von Neumann–Morgenstern preference orderings over lotteries, requiring, in addition, that they are the owners of probabilistic beliefs satisfying the Kolmogorov axioms, and that they are capable of processing these elements adequately and fruitfully. Before turning to a logical analysis of the principle of expected utility maximisation, I will give an informal overview of three elements of decision-theoretic explanations: preferences, expectations and mathematical maximisation skills.

An agent has a set of actions which, depending on what nature plays—or, in game theory, depending on what her opponents play—results in particular outcomes. The set of outcomes, however, is not the set of objects that the agent has to compare to determine what ends she prefers. Her preference ordering ranges over lotteries, that is, over probability measures on the set of outcomes.[10] If, for instance, by performing one action an agent would order stew, another kale, and still another pancakes, to be an agent with preferences in the sense required by rationality as expected utility maximisation she would not only need to have an opinion about whether she prefers stew to pancakes or kale, but also need to make up her mind about whether

[9] Weber, op. cit. 12–13 (original German).

[10] The preference ordering ranges over uncountably many alternatives, but to uncover an agent's utility functions and probabilistic expectations, only finitely many questions would have to be asked to her. For a textbook treatment of this fact, see Roger Myerson, *Game Theory: Analysis of Conflict* (Cambridge, Mass.: Harvard University Press, 1991), 12–17.

she prefers pancakes to the lottery that assigns 0.9 probability to stew and 0.1 to kale. Such human desires are perhaps a little artificial, but they are at the same time very fine-grained. To say that an agent prefers pancakes to a lottery assigning only 0.1 chance to kale is to say that she dislikes kale very much. If you give her the most preferred meal almost with certainty, and the least preferred meal with only a small probability, she will still opt for the second best option, pancakes; she just does not want to run a risk of 0.1 probability of ending up with kale.

Nor is that all, for as long as the preference ordering over lotteries does not satisfy the von Neumann–Morgenstern axioms, there is no room for expected utility maximisation to be effective as a principle of rationality. These axioms make postulations such as that no two alternatives are equally preferable and that all alternatives can be compared, and they also make it formally possible to represent the preference ordering in terms of a utility function in such a way that the intervals between the utility values truly reflect the degree of an agent's preferences.[11]

To be an expected utility maximiser, utility is not enough; you also need to have expectations. Formally, expectations are probability measures over the outcomes. The Kolmogorov axioms ensure, most importantly, that adding the probabilities of two disjoint events yields the probability of their union (in logical terms, their disjunction), and a logical consequence of this is that agents with Kolmogorov beliefs are logically omniscient.[12] Moreover, agents with Kolmogorov beliefs assign probability one to all propositional tautologies, in close correspondence to the work done by the rule of necessitation in modal epistemic logic.[13] If the beliefs of some agent are given by P, then for every tautology from classical propositional logic φ it holds that $\mathsf{P}(\varphi) = 1$.

Yet as long as an agent cannot process her preferences and probabilistic beliefs, she cannot be an expected utility maximiser. She has to be able to solve the maximisation problem that corresponds to her utility function and Kolmogorov expectations to compute the optimal action given her beliefs and desires. At first sight, it may seem that the capacity to accomplish such feats derives directly from her logical omniscience and her knowledge of all propositional tautologies, but this impression would be incorrect. Kolmogorov axioms necessary, but we also need numerous axioms that figured in the proofs of several epistemic characterisation results in order

[11] This is the content of Theorem A.1 (see the Appendix A). These requirements may seem to be hardly ever satisfied in practice, and a plethora of experiments in psychology have attempted to establish exactly that. See the relevant classic essays collected in Paul Moser (ed.), *Rationality in Action: Contemporary Approaches* (Cambridge: Cambridge University Press, 1996). Yet there is nothing intrinsically problematic with the idea that at least in some cases strategic agents actually do form von Neumann–Morgenstern preference orderings, and rather than criticising expected utility maximisation and its presuppositions in the traditional way, I here examine how far to go under the assumption that they have real bite at least in some cases.

[12] If $\mathsf{P}(\varphi) = 1$ and $\mathsf{P}(\varphi \to \psi) = 1$, then it is easy to establish that $\mathsf{P}(\psi) = \mathsf{P}(\varphi \to \psi) - \mathsf{P}(\neg\varphi) + \mathsf{P}(\neg\varphi \cap \psi) = 1$ using the Kolmogorov axioms and the observation that $\mathsf{P}(\varphi \to \psi) = \mathsf{P}(\neg\varphi \vee \psi)$.

[13] Joseph Halpern, 'The Relationship between Knowledge, Belief, and Certainty', *Annals of Mathematics and Artificial Intelligence*, 4 (1991), 301–322 contains a general result relating the logic of certainty (probability one belief) to KD45.

to allow agents to solve linear programming problems involving linear inequalities.[14]

4.1.2 Decision Theory

4.1.2.1 Explanatory Use

This determines the elements of rationality as expected utility maximisation, but it does not fix the logical form of statements involving rationality, and different readings are possible. To start with, an application of decision theory to explain the performance by agent S of action a in choice situation C takes the form:

By performing action a in choice situation C, agent S maximised her expected utility.

Extending the traditional methods of microeconomics to give an explanation of why people marry, Gary Becker, for instance, argues that people marry because they expect this to yield them higher utility than remaining single.[15] Simple though this sounds, Becker devotes a great deal of mathematical ingenuity to making the argument work, especially by carefully distinguishing all factors entering into the utility functions of the marriage partners. Yet, ultimately the decision-theoretic explanation of the fact that, say, Jane marries Bingley is the statement that by marrying him she maximised her expected utility.

Several readings of the rationality principle can be distinguished, however. To say that agent S maximised her expected utility by performing action a in choice situation C may be to make an existential or a universal statement. On the *existential* reading, this statement entails that there is some decision-theoretic model for choice situation C (that is, a decision matrix D); that there is some preference ordering over possible outcomes of her possible actions satisfying the von Neumann–Morgenstern

---

[14] A case in point can be found in the epistemic characterisation of the Nash equilibrium. True, given the axioms for linear inequalities the maximisation problem is solved automatically. But it is conceptually clarifying to separate mathematical skills from logical skills, because it shows that if, for whatever reasons, you wanted to study systems in which the logical force of the agents is weakened, you would still need to make sure that they have sufficient mathematical skills. See, e.g., Mikaël Cozic, 'Impossible States at Work: Logical Omniscience and Rational Choice', in R. Topol and B. Walliser (eds.), *Cognitive Economics: New Trends* (Amsterdam: Elsevier, 2007), 47–68. Another reason to set apart logical and mathematical capacities is that in the last decade computer scientists have gained deeper understanding of the computational complexity of solving decision processes. Here it is fruitful not to see maximisation problems as merely requiring logical ability. See Vincent Conitzer and Tuomas Sandholm, 'New Complexity Results about Nash Equilibria', *Games and Economic Behavior*, 63 (2008), 621–641 and references therein. To illustrate the relevance of this, note that as early as 1985 it was suggested that bounds to the complexity of the strategies may result in cooperation in the iterated Prisoner's Dilemma. See A. Neyman, 'Bounded Complexity Justifies Cooperation in the Finitely Repeated Prisoners' Dilemma', *Economics Letters*, 19 (1985), 227–229.

[15] *The Economic Approach to Human Behavior* (Chicago: University of Chicago Press, 1976).

Table 4.1

$\mathrm{RCT}(D,C)$	D is a decision-theoretic model for C
$\mathrm{Ut}(S,C,u)$	u is S's utility function in C
$\mathrm{VNM}(u)$	u satisfies the von Neumann–Morgenstern axioms
$\mathrm{Prob}(S,C,\mathsf{P})$	P is S's expectations in C
$\mathrm{Kolm}(\mathsf{P})$	P satisfies the Kolmogorov axioms
$\mathrm{Perf}(S,C,a)$	Action a is performable by S in C
$\mathrm{Choose}(S,C,a)$	Action a is chosen by S in C
$\mathrm{Max}(D,u,\mathsf{P},a)$	Action a solves the maximisation problem of u and P in D

axioms; and that the agent has beliefs satisfying the Kolmogorov axioms. Summing up with abbreviations as shown in Table 4.1,

$$\exists D \exists u \exists \mathsf{P}(\mathrm{RCT}(D,C) \wedge \mathrm{Ut}(S,C,u) \wedge \mathrm{VNM}(u) \wedge \mathrm{Prob}(S,C,\mathsf{P}) \wedge \mathrm{Kolm}(\mathsf{P}) \wedge \\ \mathrm{Perf}(S,C,a) \wedge \mathrm{Max}(D,u,\mathsf{P},a)),$$

which I abbreviate by $\mathrm{MEU}_{\exists}(S,C,a)$.[16] Using the same abbreviations and formalism, the *universal* reading is the statement

$$\forall D \forall u \forall \mathsf{P}((\mathrm{RCT}(D,C) \wedge \mathrm{Ut}(S,D,u) \wedge \mathrm{VNM}(u) \wedge \mathrm{Prob}(S,D,\mathsf{P}) \wedge \mathrm{Kolm}(\mathsf{P}) \wedge \\ \mathrm{Perf}(S,C,a)) \rightarrow \mathrm{Max}(D,u,\mathsf{P},a)),$$

which is abbreviated by $\mathrm{MEU}_{\forall}(S,C,a)$.

The universal reading is the weaker of the two. It does not claim that there exist utilities and probabilities, but only that action a solves the maximisation problem if such utilities and probabilities exist. This makes it less plausible as a rendering of expected utility maximisation, because it entails, for instance, that agents maximise their expected utility even in cases in which they are motivated by completely different kinds of reasons—with a false antecedent, the implication is vacuously true. Under this analysis, an agent performing an action because of tradition in the Weberian sense of the word would be maximising his or her expected utility. This is why I will ignore the universal reading in much of the remainder of this book.

More general statements about expected utility maximisation easily fit the framework by binding the relevant variables. The statement that

In choice situation C, agent S maximised her expected utility

---

[16] The clause $\mathrm{RCT}(D,C)$ states that the decision-theoretic model D represents the choice situation C. The analysis could be made more precise by specifying exactly the structure of D, and by making explicit how D and u relate. The same is true for the game-theoretic analysis below. For the purposes of my critique, however, the level of detail of the proposed analysis is sufficient. Furthermore, a distinction could be made between available actions and actions the agent believes to be available. In Chapter 1, however, I argued that in order for a decision-theoretic (or game-theoretic) model to function properly, these two sets of actions have to coincide. Cf., however, Jaakko Hintikka, *The Principles of Mathematics Revisited* (Cambridge: Cambridge University Press, 1996), 214–215.

is analysed by binding the variable a in $\text{MEU}_\exists(S,C,a)$ or $\text{MEU}_\forall(S,C,a)$ with existential or universal quantifiers, thus giving rise to the versions $\exists a\text{MEU}_\exists(S,C,a)$, $\exists a\text{MEU}_\forall(S,C,a)$, $\forall a\text{MEU}_\exists(S,C,a)$, and $\forall a\text{MEU}_\forall(S,C,a)$. These four clauses, in turn, can be used to analyse still more general statements to the effect that

> Agent S is an expected utility maximiser

as

$$\forall C(\text{MEU}(S,C)),$$

where MEU ranges over these four clauses.[17]

What do the four central clauses mean in the first place? The first one, $\exists a\text{MEU}_\exists(S,C,a)$, amounts to the conjunction of the following four claims.

1. There exists a decision-theoretic model D of choice situation C specifying the actions available to S as well as the possible consequences of these actions.
2. There exists a von Neumann–Morgenstern utility function u representing S's preferences over the possible consequences of her actions.
3. There exists a Kolmogorov probability measure P representing S's beliefs about which consequences result from which actions.
4. Action a is performable by S in choice situation C, and maximises S's expected utility given u and P.

In short, the proposition $\exists a\text{MEU}_\exists(S,C,a)$ expresses a contingent fact about the biography of agent S. As with earlier argumentation on the general form of epistemic characterisations, this clause entails that model D captures choice situation C in the way the agent represents it. It mirrors the way agent S sees C, so to speak. If S believes that she has to choose between five actions, but model D only specifies three, or if S has a different utility function or probability measure than u or P, or if instead of acting on a principle of maximisation S acted on a principle of rationality such as maximin or minimax regret, then the explanation fails to capture the reasons S had to perform a.

The same reasoning applies to the second reading, $\exists a\text{MEU}_\forall(S,C,a)$. Even though this proposition does not entail that agent S possesses preference orderings or probabilistic beliefs, it still expresses a contingent fact about S's performance of some action.

The third proposition, $\forall a\text{MEU}_\exists(S,C,a)$, states that all actions performed by S maximise expected utility given certain expectations and utility functions.[18] Assuming that S performs actions rather than not, the proposition expresses the rather sweeping claim that for any action performed by S, a decision problem was framed and a maximisation problem solved. Clearly, this claim is not necessarily true. As I have shown, Weberian traditional action is excluded here, as are actions resulting

[17] Only binding the free variable C with a universal quantifier makes sense here. The existential reading would state that an agent faced a decision problem, and that is quite different from expressing the fact that she is rational in some sense.

[18] Because the universal quantifier precedes the existential one, the utility function and the probabilistic beliefs may depend on the action.

from a Millian weighing of 'necessaries, conveniences, and luxuries', on the one hand, and 'labour and physical self-denial', on the other—at least as long as one maintains that such weighing does not necessarily involve Kolmogorov's as well as von Neumann and Morgenstern's axioms.

Finally, the proposition $\forall a\text{MEU}_{\forall}(S,C,a)$ expresses the fact that whatever action S performs, it always maximises expected utility provided that S is at that point in time the owner of a utility function and probabilistic expectations. This may seem trivially true, because why would S take pains to form a preference ordering satisfying the von Neumann–Morgenstern axioms as well as a probability measure satisfying the Kolmogorov axioms without solving the corresponding maximisation problem? For all we know, however, it may be too computationally complex to solve the problem, time may not permit it, or S may decide for whatever other reasons not to use the information of her probability measure and play maximin or minimax regret, or even adopt a boundedly rational principle such as satisficing. While that is admittedly unlikely, it is not conceptually impossible. That is why rather than making a vacuously true claim, the fourth proposition expresses a contingent biographical fact about S as much as the other three.

To summarise, decision-theoretic explanations of an agent's performing some action entail that she has to be the owner of a preference ordering or utility function over all possible outcomes satisfying the von Neumann–Morgenstern axioms; that she has probabilistic beliefs over all possible outcomes satisfying the Kolmogorov axioms; and she has the mathematical skills to solve the maximisation problem corresponding to her preferences and beliefs. Four ways to analyse the logical form of such an explanation are in principle open, but the most plausible one—and certainly the one that features in such applications of decision theory as that of marriage put forward by Becker—is, I believe, the purely existential $\exists a\text{MEU}_{\exists}(S,C,a)$. It is this one that plays a role in the remainder of this chapter.

4.1.2.2 Normative Use

Decision theory is not only used to explain, but also to advise.[19] To see how, consider an example from medicine. Crucially different treatments for breast cancer have been developed in the last 50 years, and it is often very difficult to deter-

---

[19] For a discussion of the normativity of rationality, see Raymon Boudon, 'La rationalité axiologique: Une notion essentielle pour l'analyse des phénomènes normatifs', *Sociologie et sociétés*, 31 (1999), 103–117, Francesco Guala, 'The Logic of Normative Falsification: Rationality and Experiments in Decision Theory ', *Journal of Economic Methodology*, 7, 59–93, Philippe Mongin, 'L'optimisation est-elle un critère de rationalité individuelle?', *Dialogue*, 33 (1994), 191–222, Oswald Schwemmer, 'Aspekte der Handlungsrationalität: Überlegungen zur historischen und dialogischen Struktur unseres Handelns', in H. Schnädelbach (ed.), *Rationalität: Philosophische Beiträge* (Frankfurt am Main: Suhrkamp, 1984), 175–197, Wolfgang Spohn, 'Wie kann die Theorie der Rationalität normativ und empirisch zugleich sein', in L. Eckensberger and U. Gähde (eds.), *Ethik und Empirie: Zum Zusammenspiel von begrifflicher Analyse und erfahrungswissenschaftlicher Forschung in der Ethik* (Frankfurt am Main: Suhrkamp, 1993), 151–196, and ibid., 'The Many Facets of Rationality', *Croatian Journal of Philosophy*, 2 (2002), 249–264.

mine which treatment a patient should choose. Simplifying the decision-theoretic approach somewhat, if chemotherapy, radiation therapy and surgery are the three options, then probabilities and utilities have to be assigned to all of their possible outcomes (ranging from full recovery to death), and the patient is advised to choose the treatment that maximises expected utility. Impressive research has contributed towards making these probabilities precise.[20]. In more abstract terms, typical decision-theoretic advice to S about the action to perform in choice situation C is:

Agent S must maximise her expected utility in choice situation C.

From the perspective of decision theory, the difference between explaining marriage and advising about therapy seems to be one of deontic modality—it is only about adding an operator to the effect that 'Agent S must'. This means that the deontic operator **O** can prefix the four different readings of expected utility maximisation. The clause $\mathbf{O}\exists a\mathrm{MEU}_{\exists}(S,C,a)$ means that agent S has to ensure that there is a correct decision-theoretic model for her choice situation, a von Neumann–Morgenstern utility function and a Kolmogorov probability function, as well as that she perform an action that solves the corresponding maximisation problem. $\mathbf{O}\exists a\mathrm{MEU}_{\forall}(S,C,a)$ requires that agent S perform an action solving the maximisation problem of any correct decision-theoretic model, utility function and probability measure. $\mathbf{O}\forall a\mathrm{MEU}_{\exists}(S,C,a)$ recommends to agent S that if there is some action she performs, she makes sure that there is a decision-theoretic model, utility function and probability measure with respect to which the action solves the maximisation problem. And $\mathbf{O}\forall a\mathrm{MEU}_{\forall}(S,C,a)$ tells agent S to ensure that given a correct decision-theoretic model, utility function and probability measure, if she performs some action, it then solves the corresponding maximisation problem.

The first reading is the strongest in that it requires that the agent develops a decision-theoretic model unconditionally. The third requires setting up such a model, too, but only on the condition that S performs an action. Under the third reading, the agent would conform to the norm even if she did not perform any action, and as this is true for the last reading, too, I will set both aside. The second reading requires S to perform an action even if she does not have utilities and probabilities, which also seems incoherent in a decision-theoretic context. The first reading is therefore adopted.

The analysis of the logical form of normative decision-theoretic statements requires more done than just prefixing an explanatory statement with deontic operator **O**, however, because quantifying in modal settings raises questions of *de dicto*and *de re* scoping.[21] Furthermore, since it is the actual choice of action that is recom-

[20] William Hait, David August and Bruce Haffty (eds.), *Expert Consultations in Breast Cancer: Critical Pathways and Clinical Decision Making* (New York: Marcel Dekker, 1999). The determination of the utilities was, quite understandably, left to the patients.

[21] Roderick Chisholm, 'Contrary-to-Duty Imperatives and Deontic Logic', *Analysis*, 24 (1963), 33–36 shows that a similar question arises in propositional deontic logic.

mended, an extra clause $\mathrm{Choose}(S,C,a)$ has to be inserted.[22] Quite clearly, the advice to choose an expected utility maximising action is *de dicto* with respect to the action, for otherwise it would be advice about a specific action that could be phrased without recourse to the principle of maximisation of expected utility. For the quantifiers ranging over decision-theoretic models, utility functions and probability measures, however, different readings can reasonably be set apart. As much as we can literally prefix the deontic operator,

$$\mathbf{O}\exists a\exists D\exists u\exists \mathsf{P}(\mathrm{RCT}(D,C)\wedge \mathrm{Ut}(S,C,u)\wedge \mathrm{VNM}(u)\wedge \mathrm{Prob}(S,C,\mathsf{P})\wedge \mathrm{Kolm}(\mathsf{P})\wedge \\ \mathrm{Perf}(S,C,a)\wedge \mathrm{Choose}(S,C,a)\wedge \mathrm{Max}(D,u,\mathsf{P},a)),$$

we can narrow the scope to obtain

$$\exists D\mathbf{O}\exists a\exists u\exists \mathsf{P}(\mathrm{RCT}(D,C)\wedge \mathrm{Ut}(S,C,u)\wedge \mathrm{VNM}(u)\wedge \mathrm{Prob}(S,C,\mathsf{P})\wedge \mathrm{Kolm}(\mathsf{P})\wedge \\ \mathrm{Perf}(S,C,a)\wedge \mathrm{Choose}(S,C,a)\wedge \mathrm{Max}(D,u,\mathsf{P},a)).$$

In addition to the performance of some action, the former requires that the agent make sure that there is a decision-theoretic model, a utility function and a probability measure satisfying the stipulated conditions. The latter, by contrast, presupposes the existence of a decision-theoretic model and requires, besides the performance of an action, that the agent set up a utility function and a probability measure. While the former may put a heavy burden on S's shoulders, what it requires is perfectly coherent. The latter reading, by contrast, seems to waver between presupposing that S has already adopted a decision-theoretic outlook and recommending that she take such a point of view. A more coherent picture is obtainable when it also presupposes the possession of a utility function and a probability measure with a deontic operator only ranging over actions:

$$\exists D\exists u\exists \mathsf{P}(\mathrm{RCT}(D,C)\wedge \mathrm{Ut}(S,C,u)\wedge \mathrm{VNM}(u)\wedge \mathrm{Prob}(S,C,\mathsf{P})\wedge \mathrm{Kolm}(\mathsf{P})\wedge \\ \mathbf{O}\exists a(\mathrm{Perf}(S,C,a)\wedge \mathrm{Choose}(S,C,a)\wedge \mathrm{Max}(D,u,\mathsf{P},a))).$$

In summary, the first, *large scope* reading recommends that S take a decision-theoretic outlook on her choice predicament (decision-theoretic model including the possible actions of nature, utility function and probability function) and then perform an action solving the corresponding maximisation problem. The third, *small scope* reading assumes that S has taken such a decision-theoretic view, and only tells her to perform an action that maximises expected utility (rather than, say, minimax regret). The second reading, finally, is set aside.[23]

---

[22] This clause was avoided in explanatory settings so as to make sure that what is to be explained (the choice of an action) does not figure in its explanation.

[23] My choice here is arbitrary to some extent. The deontic operator could even be appended to $\mathrm{Choose}(S,C,a)$ only. This very strict reading would entail the existence of a maximising action and recommend its performance only. In what follows, what is true of the small scope reading is equally true of the strict reading.

4.1.3 Game Theory

4.1.3.1 Explanatory Use

This concludes the investigations of the logical form of decision-theoretic explanations and prescriptions. In preparation for a similar analysis of the logical form of the theory of games, let me turn to a typical explanatory application of game theory—advertisement. Advertising is costly, but without advertisements sellers sell less. Economists have developed game-theoretic models to explain the decisions that sellers make about how much to spend on advertising as well as the fact that sellers do not all set the same price. Gerard Butters, for instance, developed a highly detailed temporal market model and showed that for large numbers of buyers and sellers a simple Nash equilibrium of advertising strategies and price distributions can be derived.[24] Sellers setting high prices advertise more intensely, because it is more difficult for them to attract buyers than it is for their lower-priced competitors. They do not overadvertise, though, because that would only lead them to incur losses. Ultimately, the explanation is that, given what his or her competitors do, a seller cannot increase expected profits by adopting a different advertising strategy.

Table 4.2

$\mathrm{GT}(\Gamma, C)$	Γ is a game-theoretic model for C
$\mathrm{Ut}(S, C, u)$	u is S's utility function in C
$\mathrm{VNM}(u)$	u satisfies the von Neumann–Morgenstern axioms
$\mathrm{Perf}(S, C, a)$	Action a is performable in C
$\mathrm{Choose}(S, C, a)$	Action a is chosen by S in C
$\mathrm{Nash}(\Gamma, u, a)$	Action a is part of a Nash equilibrium for Γ with utility function u

An application of game theory that explains the performance by agent S of action a in choice situation C takes the form:

By performing action a in choice situation C, agent S played a Nash equilibrium strategy,

where reference is made to one standard ingredient of game theory: the solution concept.[25] As the arguments against the universal reading of expected utility maximisation apply to game theory, too, I only give the obvious analogue of the purely existential reading $\exists a \mathrm{MEU}_{\exists}(S, C, a)$,

$$\exists \Gamma \exists u (\mathrm{GT}(\Gamma, C) \wedge \mathrm{Ut}(S, C, u) \wedge \mathrm{VNM}(u) \wedge \mathrm{Perf}(S, C, a) \wedge \mathrm{Nash}(\Gamma, u, a)),$$

with notation as shown in Table 4.2. This amounts to three claims:

---

[24] 'Equilibrium Distributions of Sales and Advertising Prices', *Review of Economic Studies*, 44 (1977), 465–491.

[25] The Nash equilibrium is used throughout as an example, but other solution concepts can be easily accommodated.

1. There exists a game-theoretic model Γ of choice situation C specifying the actions available to S as well as the possible consequences of these actions.
2. There exists a von Neumann–Morgenstern utility function u representing S's preferences over the possible consequences of her actions.
3. Action a is performable by S in choice situation C, and is part of a Nash equilibrium.

Game-theoretic explanations differ from decision-theoretic explanations in four respects. First, the agent's beliefs do not figure in the explanation of her actions. Second, the principle of rationality (maximisation of expected utility) has been substituted by a game-theoretic solution concept (Nash equilibrium, or any other solution concept for that matter). Third, the model Γ of the choice situation is different, because the result of the agent's choice of action is determined by the choices of other agents who, in contrast to decision theory, have preferences about the outcome of the game; it is a situation of strategic conflict. Fourth, a game-theoretic explanation of S's choice in C is an explanation of the actions of her opponents in C as well. In fact, the logical form of the explanation could be revealed by writing, for a two-person game,

$$\exists\Gamma\exists u_S\exists u_T(\mathrm{GT}(\Gamma,C)\wedge\mathrm{Ut}(S,C,u_S)\wedge\mathrm{VNM}(u_S)\wedge\mathrm{Ut}(T,C,u_T)\wedge\mathrm{VNM}(u_T)\wedge \\ \mathrm{Perf}(S,C,a)\wedge\mathrm{Perf}(T,C,b)\wedge\mathrm{Nash}(\Gamma,u_S,u_T,a,b)),$$

thus making it clear that the Nash condition in fact entails that S's opponent T plays his part of the respective Nash equilibrium.

For the explanation to be adequate, it has to mirror the way agent S views the choice situation. For decision-theoretic explanations, I have argued that this means that the utility and probability functions as well as the rationality principle specified by the theorist have to be those on which S acted. As game-theoretic explanations treat the actions of all players of a game at one fell swoop, the game has to mirror not only the way S views her choice predicament, but also the way her opponents view their choice predicaments. Modelling only one choice predicament, the game describes a situation where all players have to agree on what the choice predicament is. The consequences of this observation cannot be overestimated.[26]

[26] An objection that could be raised to this analysis is that it seems to ignore games with *incomplete information* where players are modelled to be uncertain about the utility functions. This objection would miss the point, though. Such games can easily be transformed into a normal form game for which the logical analysis proffered here is unproblematic. More importantly, if player S is incompletely informed about the utility function of T, then this is modelled by having the game Γ start with some probability measure on a number of subgames in which the utilities of T are different. Player S does not know in which subgame of Γ she is; player T does know that. In order for Γ to figure in an explanation of the actions of S and T, they still have to carve up their choice situation in exactly the way of Γ. In particular, player S is informed about the utility structure of Γ, even the part of T. Accordingly, S concurs with T as well as with the theorist in the assignment of T's utility values to all leaves of the game tree. The incompleteness of S's information about T's utility function enters in once the game has started. In other words, because she knows what the game looks like, player S knows the kinds of incomplete information about T's utilities she will encounter during game-play.

First, the space of options that S believes she has must concur with what Γ represents. If S believes that she has to choose between five actions, but Γ only specifies three, then the explanation fails to capture the reasons S had for performing the action. The same holds for the utility function. Second, the space of options the opponents of S believe they have ought to concur with Γ, because otherwise the explanation does not capture their reasons for performing their actions. The same holds for their utility functions. Third, agent S has to believe that the space of her opponents' options is exactly the one specified by the model. If S believes that, say, her opponent T has five options where the model only specifies two, then agent S's reasons for performing an action are not adequately captured. Because of the 'one fell swoop' character of game-theoretic explanations, this entails that S has to see the space of options of T as T himself sees it. The same holds for T's utility function.[27] Ultimately, to make a game theorist's explanation of an agent's action adequate, the game theorist and the players of the game have to agree on what the choice situation looks like.

What is the explanatory role of the solution concept? Comparing decision-theoretic and game-theoretic explanations, it seems as if the solution concept replaces both the probability measure and the rationality principle. But what does that mean exactly? One approach to account for the role of the solution concept uses evolutionary game theory, where certain solution concepts from non-cooperative game theory can be seen as stable equilibria in dynamical systems of agents playing games repeatedly over time.[28] These agents may be seen as readjusting their beliefs over and over again, and to be acting on the principle of expected utility maximisation. This account is not very useful for current purposes, however, because there is no agent here who is repeatedly playing games. Rather, the choice predicament of the agent is a one-time event. A second, more adequate way to think of the explanatory role of solution concepts invokes epistemic characterisation theorems. I will show how below. First, I will turn to the normative use of game theory.

4.1.3.2 Normative Use

There is more to game theory, however, because while this straightens out the logical form of game theory as a descriptive theory, game theory is used normatively as well. As a case in point, suppose a car manufacturer wishes to enter a new market and hires a consultant to give advice on what price to set. Then, Charles Holt and Alvin Roth write—in an article in the *Proceedings of the National Academy of Sciences* highlighting John Nash's 1950 two-page article on the equilibrium in

---

[27] It is not necessary that the spaces of options and the utility functions be commonly known. Something like that could be the case, and depending on whether it is the case or not, one or other solution concept is more or less appropriate in the explanation (as the epistemic characterisation theorems reveal). Yet for Γ just to model the choice situation C there need not be any higher level beliefs about the epistemic states of the other players.

[28] See, e.g., Larry Samuelson, *Evolutionary Games and Equilibrium Selection* (Cambridge, Mass.: MIT Press, 1997) focuses on this topic in particular. I will return to this briefly below.

the same journal—the consultant had better advise a strategy that is part of a Nash equilibrium.[29] In more abstract terms, a typical instance of game-theoretic advice is:

Agent S must play a Nash equilibrium strategy in choice situation C,

and given the work on decision theory, it is now easy to obtain a large scope reading

$$\mathbf{O}\exists\Gamma\exists u\exists a(\mathrm{GT}(\Gamma,C)\wedge\mathrm{Ut}(S,C,u)\wedge\mathrm{VNM}(u)\wedge\mathrm{Perf}(S,C,a)\wedge\mathrm{Choose}(S,C,a)\wedge \mathrm{Nash}(\Gamma,u,a)),$$

and a small scope reading,

$$\exists\Gamma\exists u(\mathrm{GT}(\Gamma,C)\wedge\mathrm{Ut}(S,C,u)\wedge\mathrm{VNM}(u)\wedge\mathbf{O}\exists a(\mathrm{Perf}(S,C,a)\wedge\mathrm{Choose}(S,C,a)\wedge \mathrm{Nash}(\Gamma,u,a))),$$

where it should be noted once again that, in contrast to the explanatory case, a clause $\mathrm{Choose}(S,C,a)$ has been inserted to ensure that the advice really concerns the choice of a certain strategy.[30] Similar to decision theory, the only significant difference between the two readings is that the large scope reading requires agent S to adopt a picture of her choice predicament which under the small scope reading she is already presupposed to have adopted. But where decision-theoretic advice can be made sense of, it is not at all clear how to interpret the prescriptive clause about the Nash equilibrium.

In the next section I will return to the explanatory use of game theory to show that it is reducible to decision theory. In the subsequent section I will set out three possible ways to interpret normative game theory, and show them to yield bad advice, advice that is reducible to decision theory, or nonsense.

---

[29] 'The Nash Equilibrium: A Perspective', *Proceedings of the National Academy of Sciences*, 101 (2004), 3999–4002. John Nash, 'Equilibrium Points in n-person Games', ibid., 36 (1950), 48–49. Other contemporary calls for a normative conception of game theory appear in Robert Aumann and Jacques Drèze, 'When All is Said and Done, How Should You Play and What Should You Expect?', CORE Discussion Paper 2005-21 (University of Louvain, 2005), and Martin Dufwenberg, et al., 'The Consistency Principle for Set-Valued Solutions and a New Direction for Prescriptive Game Theory', *Mathematical Methods of Operations Research*, 54 (2001), 119–131. The advice is always directed at a single agent, as recommending a Nash equilibrium to a group of players is often just wrong. This point will be reinforced below in Section 4.3.1. Moreover, what is meant by normative here is something truly deontic or prescriptive ('Keep left!'). Game-theoretic explanations of conventions or efficient norms are not the target of the present argument ('In Great Britain the norm is to keep left'). See, e.g., Peter Vanderschraaf, 'Convention as Correlated Equilibrium,' *Erkenntnis*, 42 (1995), 65–87, and Martin van Hees, 'Liberalism, Efficiency, and Stability: Some Possibility Results', *Journal of Economic Theory*, 88 (1999), 294–309.

[30] In an alternative reading the deontic operator only prefixes the statement $\mathrm{Choose}(S,C,a)$.

4.2 Game Theory as an Explanatory Theory

The analysis of the logical form of game-theoretic explanations and normative statements makes it possible to defend two claims. In this section, I will argue that game-theoretic explanations of human strategic behaviour can be reduced to decision-theoretic explanation. In the next section, I will argue that game-theoretic advice is sometimes bad advice, sometimes reducible to decision theory, and in all other cases just plain nonsense.

4.2.1 The Reduction

Epistemic characterisation theorems provide sufficient conditions under which solution concepts give an adequate description of the outcome of a game, as they are implications of which the antecedent specifies conditions on the beliefs, desires and rationality principles of the players, and of which the consequent states that the outcome of game-play under these conditions satisfies the relevant solution concept. An essential component of the argument in favour of reduction is that epistemic characterisation theorems can and should be used to justify the game theorist's application of some solution concept in a purported explanation of strategic interaction. If, for instance, a game theorist uses the concept of iterated strict dominance to explain the behaviour of some agent, then he or she ought to be understood as implicitly stating that the relevant epistemic preconditions are satisfied, and that there is common true belief about rationality and utility in that choice situation.

One may find this rather rigid, but my argument is that there is no alternative plausible way to understand how, in fact, game-theoretic explanations explain. Any other way to make game theory conform to the belief–desire format of action explanation investigated in Chapter 1 would be ad hoc. Silly alternatives to make players' beliefs explicit, for instance, would say that agents may *blunder* into a Nash equilibrium, that giving beliefs is often practically impossible and not even always necessary. Another line of thought would justify the use of, for instance, the Nash equilibrium by showing that certain dynamics (the replicator dynamics) obtain that relate to the Nash equilibrium (via Lyapunov stable sets).[31] But the problem with all these proposals is that they do not fully specify the motivational factors that played a role in the performance of the action one wishes to explain. Omitting beliefs is not a serious option, because desires and rationality principles are only in rare, basic cases sufficient. But introducing dynamics is not an option either, because they are forces on populations rather than individual motivational factors for agents playing games only once. If the game theorist is to answer the question what are the beliefs of the players whose behaviour he or she explains, then the only general plausible way is to say that their beliefs are what the epistemic characterisation theorems say they are. For if they were different, the game theorist would give an incorrect explana-

[31] Larry Samuelson, op. cit.

tion. Epistemic characterisation theorems are the canonical purveyors of the beliefs of the players.[32] It is now clear that we can turn a game-theoretic explanation into a decision-theoretic one. Given some game-theoretic model of agent S's behaviour, a decision-theoretic model is developed in the following way. Agent S's available actions and her utility function are what they were in the game-theoretic model. Agent S's probability measure in the decision-theoretic model is the one described by the epistemic characterisation theorem of the solution concept used in the game-theoretic model. Finally, the principle of rationality of the decision-theoretic model is expected utility maximisation; and that's all.[33] This shows that game theory can be reduced to decision theory.

4.2.2 The Ban on Exogenous Information

This perspective on game-theoretic modelling has far-reaching consequences. Decision theorists and game theorists alike explain actions in terms of reasons (beliefs, desires, rationality), and although in game theory only the desires are explicitly given in the explanation, epistemic characterisation theorems reveal that the underlying explanatory format is the same. Game theory is decision theory in disguise, or so it seems.

4.2.2.1 A Narrow Epistemology

The similarity is highly deceptive, though. They are similar in that even though they have to defend a claim $\mathrm{Ut}(S,C,u)$ to the effect that S possesses u as a utility function in C, they do not need to explain such things as why S has the preferences she has and whether they are reasonable. They have to demonstrate that $\mathrm{Ut}(S,C,u)$ holds, but not why it holds. With respect to the belief component $\mathrm{Prob}(S,C,\mathsf{P})$ of action explanation, however, decision and game theory are crucially dissimilar. Decision theorists as well as game theorists need to show that S possesses certain beliefs.

---

[32] Epistemic characterisation theorems typically have a converse direction to the effect that for any outcome of game-play satisfying the solution concept there exists an epistemic situation satisfying the conditions from the epistemic characterisation theorem. At first sight, the validity of the converse direction seems to be necessary if epistemic characterisation theorems are to furnish beliefs canonically. This appearance is misleading, though. Suppose that for some solution concept the converse does not hold, and that some outcome $(a_1,\dots,a_N)$ cannot result in an epistemic situation satisfying the relevant conditions. Then, if a game theorist explained i's behaviour by noting that a_i is part of an outcome satisfying the solution concept, he or she would be unable to give an explanation of a_i following the belief–desire framework (or the particular solution concept should be abandoned altogether).

[33] Anticipating the discussion later, a similar move would not work to reinstall normative interpretations of game theory. The idea would be to take it that a game-theoretic advisor implicitly assumes that the beliefs of the players are such as the relevant epistemic characterisation result specifies. Then, however, the point of the advice would disappear, because expected utility maximisation would suffice.

For a decision theorist, this is captured directly in the requirement to the effect that $\text{Prob}(S,C,\mathsf{P})$; for a game theorist, this means that he or she is committed to ascribing to S the beliefs stipulated by the corresponding epistemic characterisation theorem, which in the end amounts just as much to establishing a claim to the effect that $\text{Prob}(S,C,\mathsf{P})$. The difference is, however, that decision theorists need not explain such things as why S has the beliefs she has, but game theorists need to do exactly that because of their implicit commitment to epistemic characterisation theorems as purveyors of beliefs. This commitment entails that these beliefs do not derive from, say, statistical observations or mere guesswork, but from such epistemic set-ups as common true belief about rationality and utility—the ban on exogenous information excludes everything else. To quote the founding fathers of game theory again:

> Every participant can determine the variables which describe his own actions but not those of the others. Nevertheless those 'alien' variables cannot, from his point of view, be described by statistical assumptions. This is because the others are guided, just as he himself, by rational principles.[34]

Whenever a game theorist explains the behaviour of some agent in truly game-theoretic terms, he or she is implicitly committed to the view that in order to form the beliefs necessary for their strategic deliberation, the players disregard all available information except that which involves the game structure and rationality.

Such an epistemic policy is implausible. It is implausible descriptively in that in many situations of strategic interaction, individuals form their beliefs on the basis of information that goes far beyond rationality and game structures; and it is implausible prescriptively in that ideal epistemic subjects never rule out any informational sources unless they have good reasons to do so—and here such reasons are absent. If I want to know whether a car seller is going to accept my offer for a second-hand car, I can try to find out how this car seller generally responds to offers that are, say, 20% below the asking price, or I can consult Internet databases detailing the prices of this particular type of car with a certain age and mileage. Of course, I could inspect the game tree and assume that the seller is rational, but there is no reason for not gathering other information as well. Game-theoretic modelling entails exactly that, however. Players disregard any information that is not derived from the game structure or rationality. So, as a descriptive endeavour, game theory is committed to ascribing a narrow epistemology to the individuals that it models.

One might object to this line of reasoning on the grounds that it presupposes that the ban on exogenous information cannot be lifted. Interestingly, however, to lift the ban is to turn exactly from game theory to decision theory. If all kinds of information are allowed to play a role in an agent's decision making process, she is no longer a genuine game-theoretic agent. With the ban, game theory is reducible to decision theory and is shown to ascribe a narrow epistemology to players of games. Without the ban, by contrast, game theory is just decision theory.

---

[34] John von Neumann and Oskar Morgenstern, *Theory of Games and Economic Behavior* (Princeton: Princeton University Press, 1944), 11.

4.2.2.2 The Correlated Equilibrium

The solution concept of correlated equilibrium could form the basis of an objection to my interpretation of von Neumann and Morgenstern's ban on exogenous data, however. Given a normal form game $\Gamma = (I, (A_i)_i, (u_i)_i)$, a *correlated equilibrium* is a probability measure P over $\prod_i A_i$ such that for any player i and strategy $b_i \in A_i$

$$\sum_{(a_1,\ldots,a_N)\in\prod_i A_i} \mathsf{P}(a_1,\ldots,a_N)u_i(a_i,a_{-i}) \geq \sum_{(a_1,\ldots,a_N)\in\prod_i A_i} \mathsf{P}(a_1,\ldots,a_N)u_i(b_i,a_{-i}).$$

The use of probability generalises mixed strategies. While mixed strategies model individuals whose behaviour depends on randomisation devices or other observable probabilistic events, the correlated equilibrium has all individuals randomise, and it may be thought that randomisation is a way to model exogenous information.

But that impression would be incorrect. Sure, the probability distribution has to come from *outside* the game. It has to be generated by such things as dice, computers or sunspots.[35] The ban on exogenous information does not exclude that, though. The ban does not forbid players from making their choice of strategy dependent on outside chance events. What it does forbid is making their beliefs about other players' behaviour depend on outside events. To form beliefs about what an opponent will do, you have to restrict yourself to information about the game, typically utility and rationality. Statistical data or information about a player's character or bad temper, by contrast, are out. As I suggested in the discussion of payoff-uncertainty, factors that have a decidedly exogenous flavour (such as uncertainty about the utility functions of opponents) are nonetheless endogenous as long as no particular information about particular opponents is involved. Payoff-uncertainty as much as randomising strategies is general, and consequently, endogenous.

Yet there is an alternative (equivalent) way to define the correlated equilibrium, more common nowadays, that may suggest a different conclusion. Define, for a probability measure P constituting a correlated equilibrium, a sample space Ω consisting of all strategy profiles of the game, and define partitions $\mathscr{P}_i$ by stipulating that two strategy profiles are members of the same cell whenever they agree on the strategy of player i. Let functions $\sigma_i : \Omega \to A_i$ single out i's choice in all possible worlds. It is clear that then

$$\sum_{\omega\in\Omega} \mathsf{P}(\omega)u_i(\sigma_i(\omega),\sigma_{-i}(\omega)) \geq \sum_{\omega\in\Omega} \mathsf{P}(\omega)u_i(\tau_i(\omega),\sigma_{-i}(\omega))$$

for any τ_i. More relevant to my purposes, however, is the converse of this result, which states that for any such sample space, probability measure and partition satisfying the inequality, the (induced) probability measure constitutes a correlated equilibrium. If, interpreted as beliefs, these probability measures can possibly be shown to have exogenous origin, the ban does not extend its grasp to the correlated equilibrium.

35 Adam Brandenburger and Eddie Dekel, 'Rationalizability and Correlated Equilibria', *Econometrica*, 55 (1987), 1392.

This second interpretation of the correlated equilibrium also remains fully in the domain of endogenous information. The beliefs may be probabilistic, but unlike the Nash equilibrium case, they can be related here to common knowledge of rationality—the personification of the ban on exogenous information.[36] In fact, Robert Aumann describes the result as showing that 'common knowledge that each player chooses a strategy that maximizes his expected utility given his information.... [entails that] the strategies chosen by the players constitute a correlated equilibrium'.[37] Far from involving exogenous information, the beliefs are built in the orthodox manner, merely on the basis of information concerning rationality and utility. The ban is still at work.

Nor is that all. Another reason why the correlated equilibrium does not constitute a problem for what has been said about the ban on exogenous information is that if the probability measure is taken, in the second definition of the solution concept, to model beliefs (rather than, as in the first version, randomisation devices) all players must have the same beliefs—there is only one probability measure. While this is of course possible, it does not constitute exogenous information. If the correlated equilibrium is to model information, then it can only model situations in which there are no differences between the players with respect to the information they have obtained. That would not be a model of genuine exogenous information that particular players have about particular opponents, however; rather, it would be a model of a general endogenous informational feature of the game. The ban on exogenous information does not exclude that. What it does exclude is players forming beliefs about particular opponents in ways that defy general description in the game model.

	2_1	2_2
1_1	(3,0)	(1,2)
1_2	(0,4)	(0,1)

Fig. 4.1

And indeed, once the *common prior assumption* that all players have one and the same probability measure is relaxed and each player gets her personal probability measure with personalised requirements, the solution concept of *subjective correlated equilibrium* arises which does allow for some exogenous information. To

[36] See Section 2.1.2.2 for a discussion of the two interpretations of mixed strategies that are paralleled by the two ways to define the correlated equilibrium, and an argument why the Nash equilibrium does not conform to the ban on exogenous information.

[37] 'Correlated Equilibrium as an Expression of Bayesian Rationality', *Econometrica*, 55 (1987), 2.

see that, consider the game shown in Figure 4.1.[38] Let Ω be a two-element sample space with player 1 assigning probability one to the world in which both players play their first strategy, and player 2 assigning equal probability to this world and to the world in which they play profile $(1_2, 2_1)$. Choosing a strictly undominated strategy, player 1 is clearly rational. Given that player 2 assigns equal probability to both of his opponent's strategies, playing 2_1 is optimal, affording him an expected utility of 2 instead of $\frac{3}{2}$. Player 2 is rational, too. The beliefs constitute a subjective correlated equilibrium.

This is certainly a very serious candidate for a model of exogenous information. It is true that it does not model the belief formation of the players, nor does it model the precise exogenous origin of their beliefs, but it does make it possible to plug in exogenously formed beliefs. Ironically, however, the concept of subjective correlated equilibrium seems to be discarded by most game theorists. Robert Aumann, for instance, notes that

> the concept of subjective correlated equilibrium places very few restrictions on the possible outcomes. To get the flavor of this, note that for *any* two-person game, not necessarily zero-sum, in which there are no weakly dominated actions, there is a subjective correlated equilibrium at which both players assign positive probability to each outcome (action pair); and there is also a subjective correlated equilibrium that gives each of the two players an expected payoff as close as you wish to his maximum possible payoff. Such results indicate that the subjective correlated equilibrium is a relatively *weak* concept, giving little information; and that while logically consistent, it involves a conceptual inconsistency between the players, which distorts and hides the conflict of interests that is the subject of game theoretic analysis.[39]

In other words, the problem with the subjective correlated equilibrium is that it does not make precise enough predictions and assigns possibly incompatible beliefs to players.

Sharing this critique, Adam Brandenburger and Eddie Dekel tried to develop a refinement of the subjective correlated equilibrium that is more convincing (and equivalent to a concept characterised in terms of common knowledge) thus aligning it to the ban on exogenous information. Aumann and Drèze, art. cit. have also attempted to do something similar.[40] It is certainly true that the subjective correlated

---

[38] The game is due to Brandenburger and Dekel, art. cit. 1394. The interpretation I consider is not theirs, though.

[39] Art. cit. 15.

[40] Adam Brandenburger and Eddie Dekel, 'Rationalizability and Correlated Equilibria', *Econometrica*, 55 (1987), 1391–1402. Robert Aumann and Jacques Drèze, 'When All is Said and Done, How Should You Play and What Should You Expect?', to be published in xxx, express similar views. The starting point in the latter paper is a view voiced by Joseph Kadane and Patrick Larkey, 'Subjective Probability and the Theory of Games," *Management Science*, 28 (1982), 113–120 to the effect that players of games should proceed on the basis of subjective probability estimates concerning their opponents' actions, and that to form such estimates disciplines such as psychology are more suitable than non-cooperative game theory. Aumann and Drèze find this view 'on its face... straightforward and reasonable', but note that Kadane and Larkey overlook the 'fundamental insight of game theory: that a rational player should take into account that *all the players are rational, and reason about each other*' (emphasis in original), which they call 'interactive rationality'. They continue that 'unlike Kadane and Larkey, we note that the demands of interactive

equilibrium allows for the modelling of exogenous information. It is equally true, however, that that is an exact reason why game theorists shun the concept or modify it. The ban on exogenous information is still at work.

4.3 Game Theory as a Normative Theory

If game-theoretic explanations are reducible to decision theory and ascribe a narrow epistemology to strategically interacting agents, then the advice that prescriptive game theory gives (to which I now turn) is bad, reducible to decision theory, or nonsensical.

4.3.1 Collective Advice

.

	2_1	2_2
1_1	(0,0)	(2,-1)
1_2	(-1,2)	(1,1)

Fig. 4.2

To start with, it is instructive to see that prescriptive game theory does not succeed in giving strategic advice to collective entities (groups of players), because it does not have the conceptual apparatus to treat such entities well. It cannot talk about common purposes in the way that social choice theory studies collective aggregation of utility functions, for instance, nor does it shed light on coalition formation when players team up with allies to cooperate, as cooperative game theory aspires to.[41] Non-cooperative game theory, that is, cannot account for the binding forces

rationality severely restrict the expectations, and go on to characterise precisely what expectations can arise under this restriction'. In other words, where Kadane and Larkey still left room for exogenous information (in fact, they claimed that the only serious source of information is exogenous), Aumann and Drèze decide against it.

[41] Cf. Michael Bacharach, *Beyond Individual Choice: Teams and Frames in Game Theory*, eds. N. Gold and R. Sugden (Princeton: Princeton University Press, 2006) developing a truly collective form of normative game theory (which is very different from traditional non-cooperative game theory, though).

which groups need to have to ensure that individuals comply with the common purpose even if their individually contributing actions are not individually—and only collectively—optimal. Of course, it is not strictly speaking excluded that the theory of games gives good advice to the groups of players of a game in some rare cases, but it is striking to see that in genuinely standard cases such as the Prisoner's Dilemma shown in Figure 4.2, it fails to single out the *social optimum*. Assuming that the players want to maximise the sum total of their utilities, the Nash equilibrium of this game (according to which the players play their first strategy) prescribes a socially suboptimal outcome. The non-equilibrium outcome reached if both players play their second strategy, by contrast, affords them highest total utility. The Nash equilibrium gives wrong advice to the players as a group.

4.3.2 Individual Advice

Let me now turn to non-cooperative game theory as a normative theory for individual players. I will show, first, that in addition to the earlier observations about scoping, three interpretations of normative game theory can be set apart. Second, I will argue that under each of them the advice is problematic.

Under the large scope reading, to say that agent S in some choice predicament C must play a strategy that is part of a Nash equilibrium (or any other solution concept) is to say that S has to model her choice predicament as a non-cooperative game and 'play Nash'. Under the small scope reading, by contrast, agent S is presupposed to have such a model in mind; she is called upon to 'play Nash' only. As much as in the explanatory setting, in a prescriptive setting the clause $\mathrm{GT}(\Gamma, C)$ entails that S is advised to take (large scope) or is presupposed to have taken (small scope) a game-theoretic view of the situation. Since more players are involved in the game, agent S's view of the situation has to agree with those of her opponents. Suppose, for instance, that S thinks of the situation as a truly game-theoretic one, but in reality there are no other individuals involved, or only individuals who think they are not playing a game, or individuals who think they are playing a different game. Then playing part of a Nash equilibrium would be ill-advised and based on a misrepresentation of the situation which S should first correct. When agent S believes she is playing chess while her opponent thinks they are playing checkers, the game theorist is wrong to tell S to play the Queen's gambit.

While the way to think about the clause $\mathrm{GT}(\Gamma, C)$ is straightforward, it is more difficult to make sense of the clause $\mathrm{Nash}(\Gamma, u, a)$ in a normative setting. A strict analogy with the explanatory case would have the outcome that to make the prescriptive claim that S must play Nash is to require that S ensure that she and her opponents will actually end up in a Nash equilibrium. Although in exceptional cases S may have such powers, this interpretation would in general extend the domain of the prescriptive to the other players as well. Advice to S would be advice to her opponents, too. An alternative to this interpretation in terms of actual outcomes is that S is only advised to make a Nash outcome probable, and another alternative is that *S* is advised to make it possible. Under each of the three interpretations (actual, probable, possible) prescriptive game theory is problematic, however. The advice to play Nash or any other solution concept makes no sense if it means that such an actual outcome has to be guaranteed or that such a possible outcome should not be excluded, and it can be reduced to decision-theoretic advice if it means that such an outcome has to be made probable.

4.3.2.1 Actuality

If agent *S* must play a strategy that is actually part of a Nash equilibrium, then the advice either presupposes that her opponents concur in playing their parts of the same Nash equilibrium, or it involves advice to her opponents as well. This is because *S* cannot on her own ensure that she will follow the advice correctly. Only in degenerate cases where, for instance, all outcomes of a game are Nash equilibria, will agent *S* be able to enforce a Nash equilibrium all by herself; in general, she is dependent on her opponents for her success.

To consider the first possibility, if advising agent *S* to play a strategy that is actually part of a Nash equilibrium involves a presupposition about her opponents, then game-theoretic advice can be reduced to decision-theoretic advice. The presupposition that the opponents of *S* concur in playing their parts of one and the same Nash equilibrium can be warranted in situations in which the game theorist advising *S* has information about the prospective behaviour of *S*'s opponents. To be precise, the theorist knows that they will play their parts of the Nash equilibrium. In that case, however, the game theorist could simply pass on this information to *S*, and give her decision-theoretic advice to maximise her utility given the information. The epistemic characterisation theorem of the Nash equilibrium entails that given knowledge of the fact that *S*'s opponents play their parts of a Nash equilibrium, it maximises *S*'s expected utility to play her part. That is, game-theoretic advice would be reduced to decision-theoretic advice.

If advising *S* to play a strategy that is actually part of a Nash equilibrium involves advice to her opponents as well (the second possibility), then game-theoretic advice is bad collective advice in disguise, or simply nonsensical. As I suggested above, to play Nash in the Prisoner's Dilemma shown in Figure 4.2 is socially suboptimal and completely unsuitable as collective advice. Moreover, directed at an individual it would be nonsensical to take the advice to *S* to play Nash as also involving advice to her opponents. If a game theorist hired by a car manufacturer advises her to set her prices in equilibrium, and adds the proviso that for the advice to be really appropriate the game theorist must first advise the manufacturer's competitors as well, the manufacturer would be better served by finding another consultant. Under the interpretation of 'play Nash' as commanding an actual Nash equilibrium, game theory yields advice that is bad, nonsensical, or reducible to decision theory.

4.3.2.2 Probability

If agent *S* must play a strategy that will probably be part of a Nash equilibrium, then the advice presupposes that *S* has a way to judge degrees of probability, that is, probabilistic beliefs about the results of her possible actions. As soon as she has such beliefs, however, a decision-theoretic prescriptive statement can be made: maximise expected utility. The epistemic characterisation theorem of the Nash equilibrium (and other solution concepts) reveals that this advice agrees with the game-theoretic advice to make a Nash equilibrium probable as long as *S*'s opponents probably concur; and it may diverge considerably from game-theoretic advice in cases where *S*'s opponents are far from playing Nash. In the first case, game-theoretic advice is reducible to decision-theoretic advice; in the second, it is bad advice—plain irrational.

One could object that the game-theoretic advice given to agent *S* to play according to the Nash equilibrium entails that *S* is advised to form beliefs according to which her opponents play according to the Nash equilibrium as well. Apart from the problems with the uniqueness of the Nash equilibrium (if there are more of them, which one to consider is underdetermined) this is unacceptable for another reason. The idea would be that if a game theorist calls upon *S* to play according to the Nash equilibrium, the *logic* of such game-theoretic advice is such that he or she also calls upon *S* to form particular kinds of beliefs. But that would be very strange. To make a comparison, the logic of promises may be that if one makes a promise, one commits oneself to perform some action. But the logic of advice to perform some action is not of the kind that also entails advice to form some belief. Advice about action and advice about belief should be kept separate. Under the interpretation of 'play Nash' as commanding a probable Nash equilibrium, game theory yields bad advice or advice that is reducible to decision theory.

4.3.2.3 Possibility

If agent *S* must play a strategy that can possibly be part of a Nash equilibrium, then the advice only tells *S* to play a strategy for which there is at least one combination of her opponents' actions with which it would form a Nash equilibrium. Her opponents need not actually play that combination, they need not even probably play it; agent *S* is only required not to exclude the possibility that a Nash equilibrium will arise. This is a nonsensical recommendation. For it to be reasonable to advise *S* not to exclude the possibility of a Nash equilibrium outcome, it should be possible to explain why that is individually good for *S*. The theorist must be able to give reasons to *S* that could motivate her to perform the action she is advised to perform. If the consultant cannot explain to the car manufacturer why it would be good to set the particular price he advises, then again the manufacturer should hire another consultant. That is, the rationale behind not excluding Nash equilibria should be made explicit. And that cannot be done.[42]

[42] A different issue is whether the fact that a certain game ends in an actual Nash equilibrium in fact entails that the players played rationally. Mathias Risse, 'What is Rational about Nash Equilibria?'

To see why not, look first at decision-theoretic advice. To explain why it is sensible to act according to a rationality principle such as minimax regret, one could refer to the fact that *S* does not like to regret things, that the regret she acquires has a much more significant disvalue than, say, the disvalue attached to not obtaining the maximum possible utility, and that consequently she should minimise the maximum possible regret. To risk-averse people, by contrast, the theorist would advise the use of maximin; and to yet other individuals he or she would directly counsel expected utility maximisation. It is impossible to explain the rationale of leaving the possibility of a Nash equilibrium open in such a way, however. For a player of a game there is nothing specifically good about performing an action that does not exclude the possibility of being, at the end of the game, in equilibrium with other actions. To tell an agent not to exclude a Nash equilibrium does not connect in any way to her interests, her ways of dealing with risks, and so on. Under the interpretation of 'play Nash' as prohibiting the player to exclude the possibility that a Nash equilibrium arises, game-theoretic advice does not make sense.

Synthese, 124 (2000), 361–384 shows it does not. My interest here is, however, the reasonableness (as opposed to rationality) of the advice.

Chapter 5
The Methodology of Game Theory

This book reports both internal and external investigations into game theory. Internally, game theory's tacit modelling assumptions were investigated carefully representing and interpreting epistemic characterisation results. Externally, game theory's connections to economic consultancy or normative advice and to descriptions and explanations of rational economic agency were studied, and it emerged that the first connection is meaningless and that the second assumes a particularly narrow epistemology on the part of the modelled individuals. This chapter continues the external study. It is devoted to a view of scientific methodology that declares economic theory, if true at all, only true 'in the abstract'. In that view, game theory does not so much study strategic interaction of economic agents reasoning on the basis of beliefs, desires and rationality principles, as relate it to a field of vaguely adumbrated 'truths-in-the-abstract'.

To start, I will look at this view in the writings of John Stuart Mill and also in more recent writings of two highly respected game theorists, Robert Aumann and Ariel Rubinstein. I will then turn to three mutually dependent research habits that are sanctioned by this view—overmathematisation, introversion and model-tinkering. While these habits are not ineluctable consequences of the true-in-the-abstract view—this view may, by all means, be found in disciplines that are entirely unmathematised—in game theory it manifests itself in these research habits. The second part of the chapter contains a case study in which I compare the Nash Equilibrium Refinement Programme with the Epistemic Programme, defending the claim that the latter is more successful than the former in reaching the goal of developing a complete characterisation of the strategic rationality of economic agents and attributing this success to the stronger adherence of the Nash Equilibrium Refinement Programme to the truth-in-the-abstract methodology.

To introduce the topic, let me take a brief look at the beginnings of game theory. Suggesting a fresh look at how to apply mathematics to social science, John von Neumann and Oskar Morgenstern's *The Theory of Games and Economic Behavior* inspired a generation of mathematicians and economists to work on many questions left open by the two authors and to use their creativity to explore the

B. de Bruin, *Explaining Games*, Synthese Library 346,
DOI 10.1007/978-1-4020-9906-9_5,

new field in greater depth.[1] Of this generation, John Nash is by far the best known. The equilibrium concept that he developed in his Ph.D. dissertation quickly became a fundamental notion in game theory, spurring an entire research programme. Although a paragon of mathematical elegance and simplicity, his equilibrium concept was considered as characterising 'unreasonable' or 'counterintuitive' outcomes of game-playing. It was the ambitious goal of the Nash Equilibrium Refinement Programme (NERP) to overcome these inadequacies and to define a complete characterisation of strategic rationality in the form of the ultimate game-theoretic solution concept, while at the same time holding to many of the original maxims underlying Nash's concept.

There is virtually no textbook on game theory that does not discuss NERP in detail, while several introductions to the subject only cover NERP.[2] The programme gained the highest official recognition in 1994, when the Nobel Prize in Economics was awarded to John Harsanyi, Reinhard Selten and John Nash himself, for 'their pioneering analysis of equilibria in the theory of non-cooperative games'.[3] Robert Aumann, a Nobel laureate, too, even declared that 'the equilibrium concept of Nash..., together with its refinements, is without doubt the single game theoretic tool that is most often applied in economics'.[4]

Nevertheless, NERP has met with serious if not devastating objections. Empirical evidence soon revealed that not only the Nash equilibrium, but also all of its proposed refinements do not score highly on empirical adequacy.[5] Moreover, alternative work in game theory suggested different, more promising ways to describe economic behaviour and to reach the key objective of developing a complete char-

---

[1] Mary Ann Dimand and Robert Dimand, *A History of Game Theory: Volume I: From the Beginnings to 1945* (London: Routledge, 1996) shed interesting light on the origins of game theory prior to von Neumann and Morgenstern's *opus magnum* (its sequel, dealing with postwar game theory, will be published in 2009). Also see E. Roy Weintraub (ed.), *Towards a History of Game Theory* (ann. supp. to *History of Political Economy*) (Durham, NC: Duke University Press, 1992).

[2] Morton Davis, *Game Theory: A Nontechnical Introduction* (1970; rev. ed. 1983; repr. 1997, Mineola: Dover), Drew Fudenberg and Jean Tirole, *Game Theory* (Cambridge, Mass.: MIT Press, 1991), Roger Myerson, *Game Theory: Analysis of Conflict* (Cambridge, Mass.: Harvard University Press, 1991), Martin Osborne, *An Introduction to Game Theory* (New York: Oxford UP, 2003), id. and Ariel Rubinstein, *A Course in Game Theory* (Cambridge: MIT Press, 1994). Cf. Eric Rasmusen, *Games and Information: An Introduction to Game Theory* (Malden: Blackwell, 1989; 4th ed. 2006).

[3] The Royal Swedish Academy of Sciences, press release, 11 October 1994, http://nobelprize.org/nobel_prizes/economics/laureates/1994/press.html (accessed 11 September 2008).

[4] 'Correlated Equilibrium as an Expression of Bayesian Rationality', *Econometrica*, 55 (1987), 1.

[5] The Nash equilibrium does not fare well empirically in games with multiple equilibria and in situations in which inexperienced players play the game only once. If individuals gain experience playing games with unique Nash equilibria, they tend to converge to playing Nash equilibrium strategies over time. Evolutionary game theory tries to explain convergence as a result of trial-and-error tactics. See Larry Samuelson, *Evolutionary Games and Equilibrium Selection* (Cambridge, Mass.: MIT Press, 1997). On experimental results see Colin Camerer, *Behavioral Game Theory* (Princeton: Princeton University Press, 2003), Andrew Colman, 'Cooperation, Psychological Game Theory, and Limitations of Rationality in Social Interaction', *Behavioral and Brain Sciences*, 26 (2003), 139–198, and Douglas Davis and Charles Holt, *Experimental Economics* (Princeton: Princeton University Press, 1993).

acterisation of strategic, interactive rationality. Insights from measure theory, topology, epistemic logic, preference logic and belief–desire psychology were used—in the Epistemic Programme—instead of introducing highly ad hoc ways to refine the Nash Equilibrium. It is the contrast between these two research programmes that will be studied in this chapter.[6]

The Epistemic Programme was not, and is not, the only response to NERP. I ignore more recent contributions from evolutionary game theory, stochastic game theory and behavioural game theory.[7] These three research programmes depart more radically from NERP than the Epistemic Programme. They do not share with NERP and the Epistemic Programme the idea that game theory's key objective is to model situations of interaction between perfectly rational economic agents. Rather, they abandon rationality altogether (in favour of mutation and selection or trial-and-error, in the case of evolutionary game theory) or study less-than-perfect forms of rationality (in behavioural and stochastic game theory). While these three research programmes are generally considered to deliver models that are empirically more adequate than those of NERP, the Epistemic Programme is arguably in the same position as NERP with regard to empirical success. The difference between NERP and the Epistemic Programme highlighted here concerns a criterion of success that is, by contrast, internal. What matters here is the extent to which the theory provides adequate models of the strategic interaction of rational agents, rather than the extent to which it succeeds in describing or predicting empirical economic phenomena. NERP and the Epistemic Programme may or may not have aimed to describe and predict real economic agents. Even if they did not, it can nevertheless be meaningfully claimed that the Epistemic Programme was more successful than NERP in terms of such internal criteria of success.[8]

Game theory has received much attention from philosophers. Action theorists and philosophers of science have scrutinised the use of game theory in such fields as ethics, economics and political philosophy, while historians and science studies scholars have examined such relations as those between game theory and the military.[9] Yet, as Francesco Guala notes, there is an urgent need for case-studies of

---

[6] Several highly respected game theorists have contributed to both programmes, and I therefore define NERP and the Epistemic Programme in terms of content and methods, not people. Furthermore, finding mathematics-driven mathematisation in the contributions of a certain author does not entail that his complete oeuvre betrays overmathematisation. This is especially true of Reinhard Selten, who has shown a deep and sincere interest in empirical work, witness, e.g., his founding in 1984 of the Experimental Economics Laboratory at the University of Bonn, Germany (the first in Europe, it is claimed). See also Reinhard Selten, 'Comment [on Aumann's 'What is Game Theory Trying to Accomplish?']', in K. Arrow and S. Hohkapohja (eds.), *Frontiers of Economics* (Oxford: Blackwell, 1987), 77–87.

[7] See, e.g., Camerer, op. cit., Samuelson, op. cit., and Jacob Goeree and Charles Holt, 'Stochastic Game Theory: For Playing Games, Not Just for Doing Theory', *Proceedings of the National Academy of Sciences*, 96 (1999), 10564–10567.

[8] Nicola Giocoli, *Modeling Rational Agents: From Interwar Economics to Early Modern Game Theory* (Cheltenham: Edward Elgar, 2003), studying the history of postwar economic theory, suggests that game theory does not aim at empirical adequacy.

[9] See, e.g., Christina Bicchieri, 'Rationality and Game Theory', in A. Mele and P. Rawling (eds.), *The Handbook of Rationality* (Oxford: Oxford University Press, 2003), 182–205, Boudewijn de

concrete game-theoretic argumentation and economic model building, and it is one of the objectives of this chapter to contribute such a case-study to the literature.[10]

5.1 Truth in the Abstract

5.1.1 The Methodology

5.1.1.1 John Stuart Mill

In his 1836 essay on political economics, John Stuart Mill distinguishes between the *a posteriori method* of science, characteristic of 'those who are called practical men' who reason upward by induction on 'specific experience', and the *a priori method*, characteristic of 'those who are called theorists' who reason downward using deduction and ratiocination on the basis of abstract pictures of reality.[11] The crux of the difference—the aim is, of course, to distinguish natural and social science—lies in the possibility of carrying out an *experimentum crucis* (decisive experiment) to settle on the correctness of a theory once and for all. Mill believes that in the natural sciences a theory can be confronted with experiments that would bring to light the falsity of the theory in a direct, clear-cut and indisputable way, and hence it makes sense to consider such a theory as an inductive a posteriori generalisation of specific observed facts. In the social sciences, however, a number of characteristics lead to serious difficulties when carrying out such decisive experiments. First, the number of forces influencing a social phenomenon is much higher than in the natural sciences, and on that account it is difficult, if not impossible, to determine what exactly it is that an experiment shows—the theory may be downright false, or only incom-

---

Bruin, 'Game Theory in Philosophy', *Topoi*, 24 (2005), 197–208, Zachary Ernst, 'Explaining the Social Contract', *British Journal for the Philosophy of Science*, 52 (2001), 1–24, Francesco Guala, 'Has Game Theory been Refuted?', *Journal of Philosophy*, 103 (2006), 239–263, Daniel Hausman, 'Testing Game Theory', *Journal of Economic Methodology*, 12 (2005), 211–223, Hans Jørgen Jacobsen, 'On the Foundations of Nash Equilibrium', *Economics and Philosophy*, 12 (1996), 67–88, Harold Kincaid, 'Formal Rationality and its Pernicious Effects on the Social Sciences', *Philosophy of the Social Sciences*, 30 (2000), 67–88, Steven Kuhn, 'Reflections on Ethics and Game Theory', *Synthese*, 141 (2004), Philip Mirowski, 'When Games Grow Deadly Serious: The Military Influence on the Evolution of Game Theory' in C. Goodwin (ed.), *Economics and National Security: A History of their Interaction* (ann. supp. to *History of Political Economy*) (Durham, NC: Duke University Press, 1991), 227–256, Ahti-Veikko Pietarinen, 'Games as Formal Tools versus Games as Explanations in Logic and Science', *Foundations of Science*, 8 (2003), 317–364, Mathias Risse, 'What is Rational about Nash Equilibria?', *Synthese*, 124 (2000), 361–384, Julius Sensat, 'Game Theory and Rational Decision', *Erkenntnis*, 47 (1998), 379–410.

[10] 'Building Economic Machines: The FCC Auctions', *Studies in History and Philosophy of Science Part A*, 32 (2001), 453–454.

[11] 'On the Definition of Political Economy; and on the Method of Philosophical Investigation in that Science', *London and Westminster Review*, 26 (1836), 1–29; citations are from the 1844 ed. repr. in ibid., *Collected Works of John Stuart Mill*, ed. J. Robson (Toronto: University of Toronto Press, 1967), 309–339. On what follows, see pp. 324, 325 and 330.

plete. Second, where an experiment can circumvent this problem, it is often difficult or impossible to repeat the experiment. And in the rare cases where a decisive experiment would in principle be repeatable, it is often impossible to vary the experiment to test its robustness, resilience and reliability. This means, Mill says, that in the social sciences it does not make much sense to build theories by generalising from experience; the social sciences can only adopt the a priori method.

That being the case, the social scientist abstracts away from the phenomenon to be explained. But since he or she has to start from observation as well, it is in this process of simplification and abstraction that the alleged deductive character of the social sciences lies. On the basis of an abstract picture of the phenomenon the social scientist proceeds by mere ratiocination (deduction and calculation) to finish with statements that 'are only true, as the common phrase is, *in the abstract*'. In the process of abstraction a number of forces, causes and other factors have been ignored with the rather trivial consequence that their influence cannot be found in the statements the scientist obtains by ratiocination. In the application of the theory, these 'disturbing causes' (the causes that, in Mill's words, 'have not fallen under the cognizance of the science') have to be included, and that practice is art, not science, using the term to refer to skilled, yet creative labour rather than to such activities as painting or sculpting. Ultimately, however, 'that which is true in the abstract, is always true in the concrete with proper *allowances*'. In other words, the difference Mill detects between social and natural sciences comes down to the existence of disturbing causes. Without such causes, decisive experiments could be carried out, repeated and varied, and results as precise as those in the natural sciences would be possible.

5.1.1.2 Robert Aumann and Ariel Rubinstein

How do contemporary social scientists (and in particular, game theorists) view the methodologies they use? While in several branches of the social sciences a genuine *Grundlagenstreit* has been fought, few game theorists seem to be worried by the interpretational questions that may be raised about their subject. Very few have published their views on these matters, and if so, almost without exception in papers whose main topic was not interpretative. Two of the exceptions are Robert Aumann and Ariel Rubinstein. To start with the latter, although the emphasis in his interpretative papers lies on questions about the interpretation of particular game-theoretic models, it is fair to read Rubinstein as promoting a true-in-the-abstract view of game theory, casting the methodology mainly as a denial of any predictive aspirations underlying the theory of games. While Rubinstein does not consider game theory as a part of pure mathematics in that it relates to the real world, he rejects the view that 'the object of game theory is to predict behaviour in the same sense as the sciences do, or indeed, that it is capable of such a function'.[12] Rather, he views the function of game theory as providing abstract insights into social interaction:

> I view game theory as an analysis of the concepts used in social reasoning when dealing with situations of conflict. It is an abstract inquiry into the function and logic of social institutions and patterns of behaviour.[13]

This does not mean that there is no connection with the real world at all. Unsurprising for the true-in-the-abstract point of view, this is, however, not part of science but rather an art:

> Modeling requires intuition, common sense, and empirical data in order to determine the relevant factors entering into the players' strategic considerations and should thus be included in the model. This requirement makes the application of game theory more an art than a mechanical algorithm.[14]

An earlier contribution to the true-in-the-abstract conception of game theory is Robert Aumann's essay on what game theory is trying to accomplish.[15] Holding to the view that the concept of truth 'does not... apply to theories' and that consequently game theory is not descriptive 'in the same sense that physics or astronomy are', Aumann takes game theory to be descriptive of rational man, '*Homo rationalis*... [,] a mythical species like the unicorn and the mermaid', claiming that such a mythology does have some relevance to ordinary human beings:

> I find it somewhat surprising that our disciplines have any relation at all to real behavior. (I hope that most readers will agree that there is indeed such a relation, that we do gain *some* insight into the behavior of *Homo sapiens* by studying *Homo rationalis*).

The idea is that game-theoretic models can be expected to represent 'certain aspects of the behavior of *Homo sapiens*' in the sense that some of the phenomena we observe 'are nicely tied together by the hypothesis that they act *as if* they were maximizing'. Although this form of instrumentalism is, if you read the text carefully, only attributed to decision-theoretic and game-theoretic models of animal behaviour, Aumann nevertheless suggests that something like this may be true in the social sciences as well. Where truth cannot be the criterion to judge game theory, Aumann states that we should take an instrumentalist stance and measure its usefulness. By usefulness he does not mean usefulness as in prediction or advice, but rather the usefulness of the 'insights' that the theory provides. Nor does the term remain entirely unclear, for apart from the instrumentalist insights that Aumann

---

[12] 'Comments on the Interpretation of Game Theory', *Econometrica*, 59 (1991), 909. Other relevant papers include 'A Subjective Perspective on the Interpretation of Economic Theory', in A. Heertje (ed.), *The Makers of Modern Economics: Volume 1* (New York: Harvester Wheatsheaf, 1993), 67–83, and ibid., 'Joseph Schumpeter Lecture: A Theorist's View of Experiments', *European Economic Review*, 45 (2001), 615–628.

[13] Ibid.

[14] Ibid. 919.

[15] 'What is Game Theory Trying to Accomplish?', in K. Arrow and S. Honkapohja (eds.), *Frontiers of Economics* (Oxford: Blackwell, 1987), 28–76. On what follows, see pp. 25, 34, 36, 39 and 41–42.

promises, he proposes to value the 'classifying function' of game theory. Game theory can be used to describe 'the situations themselves rather than the behavior of the participants in them' such as when we distinguish between non-cooperative and cooperative games, or between normal form and extensive games.[16]

Finally, Aumann wishes to defend the view that game theory has to be seen as art:

> The distinction between the common conceptions of science and art is in any case not sharp; perhaps our disciplines are somewhere in between. Much of art portrays the artist's subjective view of the world; art is successful when the view expressed by the artist finds an echo in the minds of his audience... For this to happen, the artist's statement must have some universality...
>
> We [game theorists] strive to make statements that, while perhaps not falsifiable, do have some universality.

Although this may not add to Aumann's true-in-the-abstract conception of game theory, I mention it in order to make clear that when Aumann writes of game theory as an art he does not try to characterise a particular view of application. It is art in the modern meaning of 'fine arts', not in the ancient sense of *techne* (craft). It is clear that Aumann's notion of art stands in stark contrast to Mill's and Rubinstein's notions.[17]

5.1.2 The Research Habits

If these are the theoretic underpinnings of the true-in-the-abstract view, what does this view lead to in practice? Before turning to the case study pitting the Nash Equilibrium Refinement Programme against the Epistemic Programme, I will briefly

[16] I doubt whether it is genuinely possible to classify situations without behaviour. A first possibility would be to describe only the actions available to the agents, their utilities, and so on; this would be a description of the game-playing situation minus the actions actually performed. Different situations could be distinguished game-theoretically along (precisely) three dimensions (for normal form games: the (number of) players, the (number of) actions, and the utility functions), and along a fourth dimension (for extensive games: tree structure). Yet no solution concept is involved, and hence it is questionable whether such classificatory applications are genuinely game-theoretic. A second possibility would be to classify games also along the dimension of solution concepts, or at least to let solution concepts play a significant role in the classification. To be a description of situations without behaviour, solution concepts would have to be used to show possible ways of behaving (e.g., asymmetries in the game, socially suboptimal but individually rational outcomes, coordination problems about multiple Nash equilibria, etc.). Reasonable as this may sound, I doubt that the second option really makes sense without a commitment to some theory of action and motivation. An outcome can only be described as socially suboptimal and individually rational if you have some conception of what it means to be a motivation for a player to perform some action rather than another, and you can only speak about coordination problems if players envisage their choice situation in a way that is sufficiently similar to the description of the game.

[17] Rubinstein, art. cit., 919 ironically attributes his characterisation of application as art to Aumann's 1987 paper. Note furthermore the remark about falsifiability.

examine a number of research habits that arise from adherence to the true-in-the-abstract view of game theory.

5.1.2.1 Overmathematisation

Many papers in game-theoretic literature have a form that does not differ much from that of papers in pure mathematics. Relations with non-mathematical questions from economics and other social sciences, as well as with psychological and philosophical insights into rational agency are ignored, while mathematical elegance and sophistication are assigned a major role. Mathematisation, here, is driven by mathematical concerns rather than by applications. This form of mathematisation is called *overmathematisation*, overemphasising mathematics as it does.

Some game theorists acknowledge the preponderance of mathematics-driven mathematisation, but only a few of them see this as a problem. Indeed, Ken Binmore conjectures that it is commonly justified by pointing out that only in such a mathematical way can game theorists focus on important matters instead of paying undue attention to irrelevancies. However, he believes that the current approach misses this aim

> [by] leaving unformalized factors which matter, but also by introducing formal requirements that cannot be defended operationally except in terms of mathematical elegance or simplicity.[18]

Mathematisation is just a consequence of applying mathematics and is not objectionable as such, and it is often, as Giorgio Israel aptly puts it, 'the highway leading to the construction of a science that participates in the truth'.[19] When issues of applications yield to purely mathematical questions, however, this highway to the truth becomes obstructed.

5.1.2.2 Introversion

Second, a true-in-the-abstract conception often goes hand in hand with a rather introverted approach to science. Internal, technical problems of game theory are assigned great importance at the cost of questions about the external relations of the theory such as its empirical relevance or its connections to psychological, conceptual insights into human strategic behaviour. Speaking about the 'validity' of the principle of expected utility maximisation, for instance, Robert Aumann states that it

> does not depend on its being an accurate description of true individual behaviour. Rather, it derives from its being the underlying postulate that pulls together most of economic the-

[18] 'Modeling Rational Players: Part I', *Economics and Philosophy*, 3 (1987), 152.

[19] 'The Science of Complexity: Epistemological Problems and Perspectives', *Science in Context*, 18 (2005), 488.

> ory... In judging utility maximisation, we must ask not 'Is it plausible?' but 'What does it tie together, where does it lead?'[20]

Aumann continues this introverted line of reasoning to defend the claim that the alternative principle of satisficing, developed by Herbert Simon, has to be rejected:

> Alternatives [to the principle of expected utility maximisation] have proved next to useless in this respect. While attractive as hypotheses, there is little theory built on them; they pull together almost nothing; they have few interesting consequences.[21]

5.1.2.3 Model-Tinkering

Another true-in-the-abstract habit that can be found in some game theoretic literature is model-tinkering, the habit of designing all kinds of mathematical niceties to tune up some model to describe the phenomenon that it should describe. An example can be found in responses to Reinhard Selten's work on the Chain-Store Paradox.[22] Selten analysed a form of market entrance involving a monopolist with, say, hamburger restaurants in twenty towns, and twenty local business people deciding whether to enter the market by opening a local hamburger joint (once they have raised enough capital, that is). According to Selten's model, the monopolist would play cooperatively rather than aggressively in all of the twenty towns, but Selten found this clearly contrasted with everyday life economic experience, writing that this game-theoretic model 'merits the name of a paradox'.[23]

To describe aggressive behaviour on the part of the monopolist instead of cooperation, a number of game theorists started modifying Selten's model by relaxing the assumption that rationality and utility structure are common knowledge among the players. With high probability the players are expected utility maximisers, but with low probability they are not.[24] This may, of course, exactly describe the epistemic

[20] Art. cit. 35.

[21] Ibid. The implicit reference is to Herbert Simon, *Reason in Human Affair* (Stanford: Stanford University Press, 1983), but an explicit bibliographic reference to Simon is lacking in Aumann's paper. This quotation serves here mainly to illustrate introversion. Yet an inconsistency is lurking in Aumann's argument. His main objection to Simon's satisficing is that it 'ties together' much less than the principle of maximisation of expected utility. Of course, satisficing does not tie together much of the theory built around the principle of expected utility maximisation, but that is hardly surprising given the fact that that theory is deeply inspired by that very principle of expected utility maximisation rather than by satisficing. To be fair, Aumann would have to compare two fully developed theories: one centred around satisficing, the other around expected utility maximisation. The problem is, however, that while the latter of these theories exists, only the skeleton of a theory around satisficing is available to date. What Aumann should want to do (to compare the merits of two theories) is impossible. Therefore, the comparison Aumann envisages is impossible to carry out.

[22] 'The Chain-Store Paradox', *Theory and Decision*, 9 (1978), 127–159. See Section 4.1.1.

[23] Ibid. 133.

[24] David Kreps, et al., 'Rational Cooperation in the Finitely Repeated Prisoners' Dilemma', *Journal of Economic Theory*, 27 (1982), 245–252, ibid. and Robert Wilson, 'Reputation and Imperfect Information', ibid. 253–279, and Paul Milgrom and John Roberts, 'Predation, Reputation, and En-

conditions of the monopolist and the local business people, but the alternative models were not developed on the basis of close inspection of these conditions. Rather, purely mathematical model-tinkering led rather neatly to slightly different models with the desired outcome (an aggressive monopolist) and it is entirely ad hoc in that it does not have any aspirations to larger generality. The desired consistency with the aggressive monopolist therefore comes at a price, because the model not only describes the actions of the player, but also the informational structure of the game-playing situation.

This is in some respects similar to the development of physical models of motion. While secondary school physics does not correctly describe the behaviour of light or fluffy objects like balloons, ad hoc models have been developed to take care of exactly these objects by, for instance, incorporating air resistance. Air resistance is the obvious analogue of the informational structure in the above game-theoretic models, but in stark contrast to game theory, physicists can demonstrate the empirical adequacy of air resistance. The ad hoc models not only get their credibility because of the fact that they describe balloons correctly, but also by the fact that the extra assumptions concerning air resistance mirror physical phenomena of which empirical investigations actually demonstrate the 'reality'. The game theorists do not show the reality of the informational structure, however. No indication is given that monopolists and business people indeed lack common knowledge of expected utility maximisation, and that they rather assign such beliefs to each other with a certain (high) level of probability. Tuning up a model with extra assumptions without showing the adequacy of these assumptions, however, is not only ad hoc, it is also mere model-tinkering.[25]

5.2 A Case Study: Refining the Nash Equilibrium

To see the true-in-the-abstract view at work, I will now turn to a comparison between the Nash Equilibrium Refinement Programme (NERP) and the Epistemic Programme. I will first make the case that NERP researchers defended the need to refine the Nash equilibrium using epistemic terms that betray assumptions very similar to those successfully formalised in the Epistemic Programme, paying attention to the Nash equilibrium itself, the subgame-perfect equilibrium, the perfect

try Deterrence', ibid. 280–312. The solution concept used here was developed by David Kreps and Robert Wilson, 'Sequential Equilibria', *Econometrica*, 50 (1982), 863–894.

[25] In all fairness it should be noted that Kreps and Wilson, 'Rational Cooperation in the Finitely Repeated Prisoners' Dilemma', 276 seem to have anticipated this kind of critique when they wrote that one may suspect that 'by cleverly choosing the nature of [some] small uncertainty..., one can get out of a game theoretic analysis whatever one wishes. We have no formal proposition of this sort to present at this time, but we certainly share these suspicions. If this is so, then the game theoretic analysis of this type of game comes down eventually to how one picks the initial incomplete information. And nothing in the *theory* of games will help one to do this'.

equilibrium and the proper equilibrium.[26] The structure of the discussion of the solution concepts is first to examine the motivation underlying the refinement given by the NERP researchers themselves. I will lay bare the implicit epistemic assumptions that NERP researchers adopt to motivate the proposed solution concept, referring to results from the Epistemic Programme that captured these assumptions more adequately. Subsequently, a historical summary of the Epistemic Programme emphasises what I have already explained in detail in Chapters 2 and 3; namely, that its formalism is application-driven rather than mathematics-driven.

5.2.1 The Nash Equilibrium Refinement Programme

5.2.1.1 The Nash Equilibrium

John Nash worked on his equilibrium concept while he was a graduate student at Princeton, publishing a paper in *Annals of Mathematics* in 1951.[27] For the historian and philosopher of science his 1950 Ph.D. dissertation is a more interesting source to consult though, because it contains a six-page discussion of what Nash calls the 'interpretation' of the equilibrium concept he proposes.[28] Nash asserts that his equilibrium is the only rational prediction of the outcome of strategic interaction:

> We proceed by investigating the question: what would be a 'rational' prediction of the behavior to be expected of rational [*sic*] playing of the game in question? By using the principles that a rational prediction should be unique, that the players should be able to deduce and make use of it, and that such knowledge on the part of each player of what to expect the others to do should not lead him to act out of conformity with the prediction, one is led to the concept of a solution defined before.[29]

The wording is already epistemic. First, the players are supposed to be able to 'deduce' the Nash equilibrium and to 'make use of it'. This clearly presupposes knowledge about the strategies available to the players as well as their utility functions, for without strategies and utility there is no equilibrium. The players have to know what game they are playing if they are to calculate the equilibrium. Second, with no extra argumentation, Nash contends that knowledge about the equilibrium constitutes knowledge about 'what to expect the others to do'. This may seem unwarranted, for why would players play equilibrium strategies? Yet it is unsurprising once you

---

[26] I refer to the primary sources from the refinement literature. Eric van Damme, *Stability and Perfection of Nash Equilibria: Second, Revised and Enlarged Edition* (Berlin: Springer, 1987) contains very similar argumentative strategies for such refinement proposals as the approachable, essential, firm, regular, and strictly and weakly proper equilibrium.

[27] 'Non-Cooperative Games', *Annals of Mathematics*, 54 (1951), 286–295.

[28] 'Non-Cooperative Games', Ph.D. diss. (Princeton University, 1950). The note is absent from the published paper with the same title. The appendix to the Ph.D. dissertation (pp. 21–26) also mentions a second 'interpretation' of the Nash equilibrium in terms of repeated game-play, foreshadowing applications in evolutionary game theory.

[29] Op. cit. 23.

realise that Nash also stipulates that the equilibrium be the 'unique' rational prediction of the outcome of a game. Under that stipulation it is true that if one has the ability to deduce the equilibrium, one has by the same token the ability to obtain knowledge about the only rational prediction of the game, and, consequently, to obtain knowledge about what to expect from others. Every player knows what her opponents will play, and she knows that they will play according to the Nash equilibrium. Given such knowledge the only expected utility maximising action for her to perform is to play his part of the Nash equilibrium. This shows that the last of Nash's desiderata is also fulfilled. Knowledge about the rational prediction will not make players act 'out of conformity with the prediction'.

NERP researchers have long accepted as a 'folk theorem' that the Nash equilibrium requires common knowledge of rationality and utility. Nash's argument does not require this, however, and it was not until the publication of several papers on the epistemic characterisation of the Nash equilibrium that this view was corrected.[30] It was shown that if all players are rational (expected utility maximisers), and know their own utility function, and know what their opponents will play, then the players will play a Nash equilibrium.[31] The epistemic characterisation theorem of the Nash equilibrium does not contain a reference to common knowledge or to knowledge about rationality—the folk theorem is false. Furthermore, the epistemic assumptions underlying the Nash equilibrium thus disclosed leave no room for a 'deduction' of the Nash equilibrium as Nash envisaged it. To play a Nash equilibrium, players have to know exactly what their opponents will choose. Such knowledge cannot in general be obtained on the basis of knowledge about the fact that the opponents are rational, because rationality frequently does not allow one to settle on a unique course of action. It will rather be obtained on the basis of exogenous, statistical information about previous game-playing situations, if such information is available and dependable. However, following von Neumann and Morgenstern's ban on exogenous information, Nash's idea was that players will 'deduce' such knowledge on the basis of information about the structure of the game only. The epistemic characterisation of the Nash equilibrium shows that this is impossible.[32]

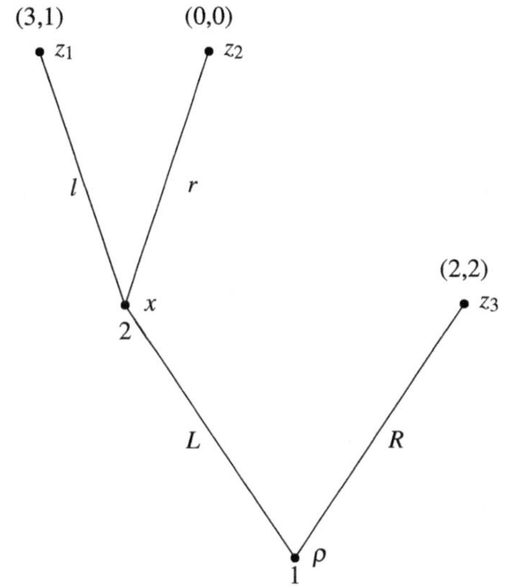

Fig. 5.1

5.2.1.2 The Subgame-Perfect Equilibrium

Reinhard Selten's 1965 paper on oligopoly with demand inertia contains the first proposal for a refined equilibrium concept.[33] Selten crafts his argument on the basis of a game that, he believes, contains a Nash equilibrium that is not 'rational', and he provides a refinement excluding exactly such alleged irrationalities. Selten's game is shown in Figure 5.1 with changed notation. It is an extensive game of perfect information. It contains two (pure) Nash equilibria, (R,r) and (L,l), the former of which, Selten states, ought not to be called 'rational':

> The equilibrium point (R,r) can only be so interpreted that player 2, prior to game-play, threatens with a decision for z_2 to drive player 1 to behave according to R. This threat, however, has to remain without effect, as long as player 2 cannot, in advance, commit himself to carry it out. Player 1 knows that player 2 will have no interest in carrying it out as soon as play has reached the point x. This reflection shows that (R,r) cannot be seen as a rational, non-cooperative solution under the assumption of a complete absence of the power to bind

---

30 Robert Aumann and Adam Brandenburger, 'Epistemic Conditions for Nash Equilibrium', *Econometrica*, 63 (1995), 1161–1180, and Adam Brandenburger, 'Knowledge and Equilibrium in Games', *Journal of Economic Perspectives*, 6 (1992), 83–101.

31 For references to papers showing adherence to the incorrect folk theorem, see Aumann and Brandenburger, art. cit.

32 Nash's second interpretation in terms of repeated game-play is not immediately excluded by the epistemic characterisation of the Nash equilibrium.

33 'Spieltheoretische Behandlung eines Oligopolmodells mit Nachfrageträgheit: Teil I: Bestimmung des dynamischen Preisgleichgewichts', *Zeitschrift für die gesamte Staatswissenschaft*, 121 (1965), 310–324.

> oneself [to a particular choice of strategy]. The possibility of such undesired cases forces us to sharpen the notion of equilibrium in an appropriate manner.[34]

The strategy profile (R,r) can only be interpreted as representing a situation in which, prior to game-play, player 2 has voiced a threat to commit himself to playing r. Nonetheless, Selten claims that this threat is not credible because r is irrational for 2 once he arrives at x. To exclude the 'irrational' strategy profile (R,r), Selten introduces a refinement of the Nash equilibrium nowadays called subgame-perfect equilibrium.[35] A *subgame-perfect equilibrium* of an extensive game is a Nash equilibrium that satisfies the additional requirement that it be a Nash equilibrium of all of its subgames. This clearly rules out the strategy profile (R,r), since in the subgame generated by x it would be better for player 2 to play l.

Selten's argument against the rationality of (R,r) only works, however, on the basis of two specific assumptions about the beliefs of the players, one of which Selten leaves fully implicit and unformalised. First, Selten has to assume that the strategies characterised by the Nash equilibrium are backed by particular beliefs held by the players. Without having beliefs about player 2, player 1 would not be in the position to evaluate player 2's announcement to play r as a threat rather than as a helpful suggestion for coordinated action. To see this it is important to examine what makes a threat a threat. If it is carried out, the threat has to be disadvantageous to the player threatened. In addition, if the threat is taken seriously by the threatened player, then this is advantageous to the one who threatens, and the threatener will not carry out the threat. For the threatened player to believe that the threatener is making a conscious threat, the one threatened has to believe that the threatener believes all this. In fact, common belief is needed.

Applied to the present case, Selten has to assume that player 1 believes that player 2 believes that r, if carried out, would be disadvantageous to player 1. This is an instance of common belief about the utility function of player 1, representing as it does what is advantageous and disadvantageous to him.[36] Selten has to assume, too, that player 1 believes that player 2 believes that if player 1 takes the threat seriously and plays R, this is advantageous to player 2. This is an instance of common belief about the utility function of player 2. Selten has to assume, moreover, that player 1 believes that player 2 believes that if player 1 believes that player 2 will carry out his threat and play r, that player 1 will then give in and play R. Since such reasoning involves the rationality of strategic choice given the beliefs and desires of player 1, this is an instance of common belief about player 1's rationality. Thus, the first assumption in order for Selten 's interpretation of (R,r) to work is that there be common belief about the utility functions of the players as well their rationality.

---

[34] Ibid. 308 (original German).

[35] Selten's own term was 'perfect equilibrium' (ibid. 308). It coincides with backward induction on games with perfect information.

[36] It is common belief up to level 2, to be precise. Larger games naturally require higher levels. See Section 3.1 to the effect that the full infinity of common belief will never be used in any finite game, but a general characterisation statement is most easily phrased using full infinite common belief.

Nor is this all. As a second presupposition, Selten has to decide on a specific form of rationality appropriate to playing extensive games by answering the question of whether the rational thing for a player to do in a certain subgame depends on the history of choices preceding the subgame. Selten explicitly adopts the view that there is no such dependence (a *history-insensitive* conception of rationality, in the terminology developed earlier) when he writes that 'if there is no power to bind oneself [to a particular choice of strategy], then the behaviour in a subgame is allowed only to depend on the structure of the subgame itself'.[37] This assumption is crucial because without it subgame-perfection would not be the right solution concept, depending as it does on the idea that there is a clear answer to what rationality entails at any decision node of a game tree.[38]

Common belief about rationality and utility, coupled with a specific, history-insensitive view of rationality in subgames, are the assumptions that inspire Selten's subgame-perfect equilibrium. It is in the Epistemic Programme that these assumptions were investigated and formalised explicitly, however. To anticipate the discussion below, a central debate in the Epistemic Programme concerns the question of whether common belief about rationality and utility entails subgame-perfect equilibria (backward induction). Subtle distinctions have been made between various forms of rationality, while the temporal dynamics of beliefs have been analysed using lexicographic probabilities and the logical theory of belief revision. To some extent the Epistemic Programme vindicates Selten's subgame-perfect equilibrium in situations of common belief about rationality and utility. It became clear, for instance, that Aumann's epistemic characterisation theorem attempts to give a rigorous proof to that effect. On the other hand, the Epistemic Programme scores better as it not only formalises the solution concept, but also the epistemic conditions, in particular various notions of history-sensitivity.

5.2.1.3 The Perfect Equilibrium

Ten years after the publication of his paper on the subgame-perfect equilibrium, Selten proposed another refinement.[39] Although subgame-perfection excludes a number of 'intuitively unreasonable' Nash equilibria, Selten argued that there is a game in which subgame-perfection does not exclude them all. The game, shown in Figure 5.2 with changed notation, has become famous under the name of *Selten's Horse*. Each of the three players can choose between two actions. There are (uncountably) infinitely many mixed Nash equilibria which are subgame-perfect as the game itself is its only subgame. They can be represented by a triple (p_1, p_2, p_3) of

---

[37] Ibid. 308.

[38] See Chapter 3 for a discussion of the views of common belief about rationality and utility developed in the Epistemic Programme, attesting that the formalism was inspired by its applications, namely, to study the behavioural consequences of particular epistemic assumptions involving the beliefs, desires and rationality of the players of extensive games.

[39] 'Reexamination of the Perfectness Concept for Equilibrium Points in Extensive Games', *International Journal of Game Theory*, 4 (1975), 25–55.

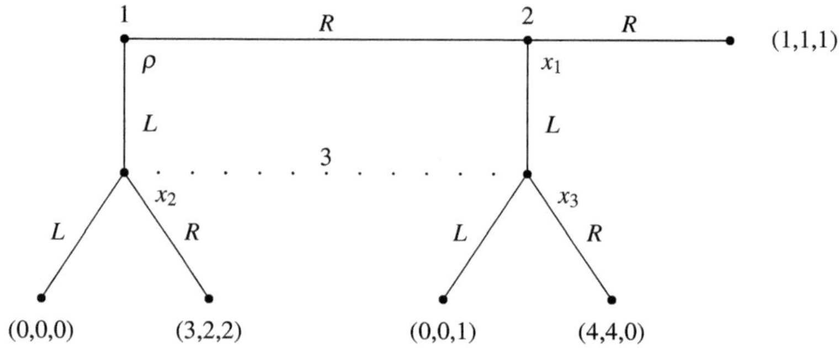

Fig. 5.2

real numbers in the unit interval, where p_i is the probability that player i plays R. One type of equilibrium is $(1,1,p_3)$ with $0 \leq p_3 \leq \frac{1}{4}$; the other type is $(0,p_2,1)$ with $\frac{1}{3} \leq p_2 \leq 1$. Selten explains that the second type of equilibrium 'cannot be regarded as reasonable':

> Now suppose that the players believe that [for example]... $(0,1,1)$... is the rational way to play the game. Is it really reasonable to believe that player 2 will choose R if [node x_1] is reached? If [player 2] believes that player 3 will choose R as prescribed by the equilibrium point, then it is better for [player 2] to select L where he will get 4 instead of R where he will get 1...
>
> Clearly, [this equilibrium] cannot be regarded as reasonable. Player 2's choices should not be guided by his payoff expectations in the whole game but by his conditional payoff expectations at [x_1]. The payoff expectation in the whole game is computed on the assumption that player 1's choice is L. At [x_1] this assumption has been shown to be wrong. Player 2 has to assume that player 1's choice was R.[40]

Again, the motivation for refinement is cast in epistemic terms. As in the paper on subgame-perfection, Selten here uses common belief about rationality and utility as well as the history-insensitive view of the temporal development of rationality. He adds a condition on the temporal development of players' beliefs, though. Selten holds that a player's beliefs at some decision node x will not be determined by beliefs the player had, at decision nodes y preceding x, about the likelihood of reaching x. Belief formation as well as rationality are, that is, history-insensitive.

The refinement of the Nash equilibrium that he proposes is called the *perfect equilibrium*. I will give a brief survey of this refinement, first to underline my claims about the use of epistemic argumentation in the motivation underlying the perfect equilibrium, and second because it gives a first illustration of mathematics-driven mathematisation in the Nash Equilibrium Refinement Programme. Given an extensive game Γ (with perfect recall), a test sequence of perturbed games is constructed. A *perturbed game* $\acute{\Gamma} = (\Gamma, \eta)$ is an ordered pair consisting of the original game and a probability function assigning positive probability to every choice any player can

---

[40] Ibid. 34.

make. A sequence of perturbed games $(\acute{\Gamma}^k)_k$, where $\acute{\Gamma}^k = (\Gamma, \eta_k)$, is a *test sequence* whenever the probabilities converge to zero for $k \to \infty$. In addition, Selten introduces the notion of *behaviour strategy* to take account of the information the probability function η_k gives to the player about the likelihood of certain outcomes of a certain perturbed (or unperturbed) game. A *limit equilibrium point* of a test sequence is defined in a straightforward manner. Finally, a Nash equilibrium of an extensive game Γ (with perfect recall) is a perfect equilibrium, in the sense that Selten defines it, if and only if there is at least one test sequence of Γ of which it is a limit equilibrium point. He proceeds to prove a theorem to the effect that every perfect equilibrium is subgame-perfect, and then makes an intricate, four-page argument to show that the undesired equilibria from the game shown in Figure 5.2 are not perfect, and that the other equilibria are, in fact, perfect; not all subgame-perfect equilibria are perfect, that is, and this shows that the perfect equilibrium is a genuine refinement of the subgame-perfect equilibrium.

5.2.1.4 The Proper Equilibrium

The reader may wonder if refining the Nash equilibrium involves nothing more than first extending its scope from normal form games to extensive games with perfect information, and then to extensive games with imperfect information. To stress that this impression would be wrong, I will conclude the survey with a discussion of Roger Myerson's critique of the Nash equilibrium for normal form games. First, he explains:

> The concept of [Nash] equilibrium...is one of the most important and elegant ideas in game theory. Unfortunately, a game can have many Nash equilibria, and some of these equilibria may be inconsistent with our intuitive notions about what should be the outcome of a game.[41]

	l	r
L	(1,1)	(0,0)
R	(0,0)	(0,0)

Fig. 5.3

On the basis of the game shown in Figure 5.3 with changed notation he continues:

41 'Refinements of the Nash Equilibrium Concept', *International Journal of Game Theory*, 7 (1978), 73.

> To see how these counterintuitive equilibria can arise, consider the game [shown in Figure 5.3]. There are two Nash equilibria in this game, (L,l) and (R,r)... It would, however, be unreasonable to predict (R,r) as the outcome of this game. If player 1 thought that there was any chance of player 2 using l, then player 1 would certainly prefer L. Thus (R,r) qualifies as an equilibrium only because Nash's definition presumes that a player will ignore all parts of the payoff matrix corresponding to opponents [*sic*] strategies which are given zero probability.[42]

Myerson's main point is that when player 1 believes (with arbitrarily small probability) that player 2 may play l, then player 1, if rational, would play L. Yet, if there is no reason why player 1 would adopt such beliefs, there is no point complaining about the Nash equilibrium (R,r). While Myerson does not state it explicitly, it is not difficult to reconstruct his argument in full. One of the reasons why player 1 would adopt such a belief is that l does not score less than r for player 2 against player 1 playing R, and l scores better against player 1 playing L. It is, in other words, a (weakly) undominated strategy, and it is a form of rationality to choose these.

The mathematical techniques Myerson uses to set up the refinement are vaguely similar to those used by Selten to develop his perfect equilibrium. First, Myerson defines an *ε-proper equilibrium* as a profile of completely mixed strategies $(\sigma_1,\ldots,\sigma_N)$ such that each player assigns much higher probability to her better choices than to her worse ones. More precisely, if it is better to play l than k given the mixed strategies of player i's opponents, then i's mixed strategy should assign a factor of $\frac{1}{\varepsilon}$ greater probability to l than to k. Myerson defines a *proper equilibrium* as the limit of a sequence of ε-proper equilibria satisfying certain straightforward mathematical conditions. He goes on to show that the set of proper equilibria is contained in the set of Nash equilibria, thus showing it to be a real refinement. Finally, a rather intricate theorem shows that every game has at least one proper equilibrium.[43]

5.2.2 *Mathematics-Driven Mathematisation in the Nash Equilibrium Refinement Programme*

Up until now I have argued that the proposed refinements of the Nash equilibrium were motivated by very specific epistemic assumptions, and that they were meant to characterise the outcome of strategic interaction between rational economic agents in situations satisfying exactly these assumptions. For Nash, the only rationale underlying the quest for the ultimate solution concept was to determine the unique, rational prediction of rational play by intelligent economic agents. Such an outcome,

---

[42] Ibid. 73–74.

[43] Results in the Epistemic Programme show that Myerson's proper equilibrium does not adequately capture the epistemic assumptions he implicitly makes. For further discussion of the form of rationality excluding weakly dominated strategies and a solution concept based on it, see Section 2.3. For an enlightening discussion and a recent alternative to the proper equilibrium, see, e.g., Geir Asheim, 'Proper Rationalizability in Lexicographic Beliefs', *International Journal of Game Theory*, 30 (2002), 452–478.

he claimed, can only be an equilibrium, and so he devised his solution concept as one in which no player would be better off if she were to choose a different strategy. Subsequent NERP researchers tried to stay as close as possible to this idea.

The treatment of the Epistemic Programme in Chapters 2 and 3 suggests, moreover, that by taking a different route it was more successful in giving such a characterisation because it provided explicit formal models of these epistemic assumptions. It tried to demonstrate systematic connections between the players' beliefs and utility functions and the game's outcome. Since the Epistemic Programme was not committed to looking for a unique outcome, a whole new realm of mathematical results was opened, relating epistemic conditions to various non-equilibrium solution concepts by means of epistemic characterisation theorems. Iterated strict dominance, for instance, arises in normal form games with common belief about rationality and game structure, and the same epistemic assumptions in extensive game-playing give rise to backward induction. The Epistemic Programme has given rise to different models relaxing these assumptions. Both the probability theory of lexicographic beliefs and the logic of belief revision have inspired precise characterisations of outcomes that arise when players adopt history-sensitive conceptions of rationality and belief, for instance, or when they are less than fully informed about their opponents' utility functions.[44]

Judging by the increased attention the Epistemic Programme has received since the mid-1990s in such highly respected journals as *Econometrica*, the *Journal of Economic Theory*, *Games and Economic Behavior*, and the *International Journal of Game Theory*, and also by the decline in the number of publications on NERP, the Epistemic Programme would seem to have won the day. The Epistemic Programme has found a home in two biennial conferences devoted solely to the study of belief in strategic interaction (Theoretical Aspects of Rationality and Knowledge (TARK) and Logic and the Foundations of Game and Decision Theory (LOFT)), and it has become a highly interdisciplinary field, attracting attention from, among others, game theorists, logicians, philosophers, probability theorists and statisticians.[45]

While it would be rash to claim that the Epistemic Programme has outrun NERP on empirical adequacy, the claim defended here is that the Epistemic Programme was more successful than NERP if you measure success by the extent to which the programmes succeeded in reaching the main research objective of providing a complete characterisation of strategic rationality. I will now consider an explanation of the differences in success in terms of overmathematisation and other true-in-the-abstract research habits.

---

[44] NERP and the Epistemic Programme have very similar aims in that both attempt to determine what players will choose under certain epistemic and rationality assumptions. They share, for instance, the conception of rationality as expected utility maximisation, and they share, as I have argued here, an interest in common belief, etc. But in NERP these assumptions were not explicitly formalised and a lot of theorising remained mathematics-driven rather than application-driven. In the Epistemic Programme, by contrast, these assumptions were formalised with highly technical, yet in the end application-driven mathematical tools.

[45] It is interesting to note that TARK started as an acronym for 'Theoretical Aspects of Reasoning about Knowledge', and changed to the current name in 1998, reflecting an increase in attention on decision and game theory.

The NERP literature is decidedly Bourbakian, with definitions preceding theorems, which in turn precede proofs.[46] Existence theorems, uniqueness results, interrelation lemmas and mathematical pleasantries like the *decentralisation* theorem for the perfect equilibrium are brought to the foreground.[47] Questions about application and empirical relevance are left untouched, as are conceptual questions about rationality, the role of beliefs and knowledge in decision-making, the specifics of common belief in strategic interaction, and the subtle differences between various forms of rationality—overmathematisation.

There is already mathematics-driven mathematisation in Nash's very idea to generalise von Neumann and Morgenstern's solution concept. Nash extends the concept to apply to games of arbitrary numbers of players (instead of two) and to games with arbitrary utility functions (instead of zero-sum games), but he does not give a conceptual or empirical reason why such generalisation would increase our understanding of economic phenomena involving rational strategic interaction. In his own words, the proposed equilibrium notion yields

> a generalization of the concept of the solution of a two-person zero-sum game [in the sense that] the set of [Nash equilibria] of a two-person zero-sum game is simply the set of all pairs of opposing 'good strategies'.[48]

Von Neumann and Morgenstern, by contrast, use conceptual and empirical argumentation to defend the opposite claim that such a generalisation is unnecessary and even misguided from the viewpoint of applications. They contend that games with more than two players are no longer non-cooperative because groups of players will team up to play in a mutually beneficial way. Moreover, in non-zero sum games the players' reasoning processes will not substantiate anything like the concept of solution that they themselves provide.[49] Nash does not address these issues at all, however. He does not explain why he thinks that to understand strategic interaction there is a need, contra von Neumann and Morgenstern, to generalise their concept. Irrespective of the question of whether von Neumann and Morgenstern were right or wrong—there is much evidence that they were, indeed, too strict—Nash's silence here indicates that his motivation was inspired not so much by questions of application as by purely mathematical questions.[50]

---

[46] See, e.g., E. Roy Weintraub and Philip Mirowski, 'The Pure and the Applied: Bourbakism Comes to Mathematical Economics', *Science in Context*, 7 (1994), 245–272.

[47] Selten, 'Reexamination'.

[48] Nash, art. cit. 286.

[49] John von Neumann and Oskar Morgenstern, *Theory of Games and Economic Behavior* (Princeton: Princeton University Press, 1944), 31–33, 220–221.

[50] Von Neumann judged Nash's result to be 'trivial... [and] just a fixed point theorem' (Sylvia Nasar, *A Beautiful Mind: A Biography of John Forbes Nash, Jr., Winner of the Nobel Prize in Economics, 1994* (New York: Simon and Schuster, 1998), 94). Note, however, that in view of von Neumann's rejection of subjective probability and references to players' beliefs, Nash's interpretation is perhaps quite radical in a sense. The biographical detail about von Neumann's life may be read, moreover, as indicating that having considered the mathematical generalisation himself, he (and Morgenstern) found that it did not yield any interesting contribution to the modelling of economic phenomena. The way Nash introduced the generalisation did not change this conviction,

A second instance of overmathematisation can be discerned in Nash's insistence on requiring the uniqueness of the outcome of a game. Hardly motivated by non-mathematical concerns, Nash shared with most other mathematicians his fondness for uniqueness proofs.[51] A simple search through the bibliographic database MathSciNet of the American Mathematical Society reveals around 15,000 publications (after 1950) with the word *unique* (or compounds thereof) in the title.[52] But where uniqueness is often justified as a desideratum within mathematical theory, it is difficult to justify in the applied context within which Nash was working; not even if the desideratum is taken to be a scientific one aspiring to the highest levels of falsifiability. Rationality does not often fix one single action, witness the famous case of Buridan's ass, which had to decide between two equally attractive hay bales, or, to take a more extreme case, games in which there are several outcomes that are equally attractive to all players.

To sum up, Nash's work betrays overmathematisation in that his mathematical modelling assumptions are not backed by conceptual or empirical insights about what the models are intended to model. By contrast, research in the Epistemic Programme started from the idea that outcomes of games had to be related to epistemic conditions involving the key motivational elements of human agency: beliefs, desires and rationality, and it succeeded in staying much closer to the original idea, which was to study the strategic interaction of rational individuals.

The first refinement of the Nash equilibrium immediately poses a challenge to the view defended in this paper, however. Selten presents the subgame-perfect equilibrium in a research report on oligopoly with demand inertia, rather than in a mathematics-driven context.[53] He takes time to introduce the reader to the relevant economic phenomena, carefully introduces a mathematical model by defending a set of equations and (on the basis of these equations) an extensive game. He observes that in this game the Nash equilibrium fails to characterise only adequate outcomes, and concludes that this shows the deficiency of the solution concept. Then, in little more than two pages, he submits a refinement of the Nash equilibrium, the subgame-perfect equilibrium. Of course there is mathematics in this paper, but this is not emphasised typographically, textually, or conceptually. The stress lies on a model of particular economic phenomena and some of its properties. That is, the motivation for refining the Nash equilibrium concept is, in this paper, ultimately driven by concrete problems of application, and one should expect that, if it is true that the different levels of success of NERP and the Epistemic Programme can be attributed to

---

because Nash did not show much awareness of economic applications, not even in the interpretative Appendix to his Ph.D. dissertation (op. cit. 21–26).

[51] Nash did allow for this insight into the sense that all strategies that receive non-zero probability in a Nash equilibrium that consists of mixed strategies are equally good given that the opponents stick to their equilibrium strategies. This draws me into a discussion about whether a probability distribution over actions (which is what a mixed strategy really is) models anything like actions (as Nash seems to suggest for repeated games), or rather beliefs (as the standard view seems to be nowadays, witness the treatment in most textbooks). See Section 2.1.

[52] Using search-term 'unique* OR unicity', and 1950–present as the year-range, 15,460 hits were found (July 10, 2010).

[53] 'Spieltheoretische Behandlung'.

NERP's overmathematisation, Selten's 1965 results come closest to corresponding findings from the Epistemic Programme.

This is in fact the case. The subgame-perfect equilibrium and the much older concept of backward induction are the same, at least concerning games with perfect information, and the epistemic characterisation theorem obtained in the Epistemic Programme for backward induction agrees with Selten's argumentation to the effect that subgame-perfect equilibria arise in situations in which there is common belief about rationality and utility (with, to be precise, a history-insensitive variant of rationality).[54]

There is a difference, though, for Selten presents a simple, mathematically elegant, but conceptually ad hoc solution to the problem he tries to solve. Simple, because it applies a preexisting definition of Nash equilibrium to subgames, and mathematically elegant because it tests for a property inductively through the game tree. It is conceptually ad hoc, however, because the connection with the conceptual motivations (credible threats, binding forces, rationality) remains very tenuous. Selten stayed with the techniques that Nash developed, without questioning whether this really suited his modelling purposes. The Epistemic Programme, by contrast, developed completely new tools to study exactly the kind of strategic interaction under the same epistemic conditions (and variants thereof).

It is striking to observe that, ten years later, Selten's work has become entirely introverted and mathematics-driven, shunning as it does all references to economic applications and strategic interaction of rational agents.[55] He starts his paper on the perfect equilibrium with some six pages of definitions, then states and proves a theorem with the help of two lemmas, and presents an argument of less than two pages for the claim that subgame-perfect equilibria are sometimes 'intuitively unreasonable'.[56] Selten then introduces the technical apparatus needed to define his refinement of the subgame-perfect equilibrium, the perfect equilibrium. A series of theorems demonstrate that in games of perfect recall all perfect equilibria are subgame-perfect, that not all subgame-perfect equilibria are perfect, that a decentralisation property holds, that all games of perfect recall have at least one perfect

[54] It would be both conceptually and historically inappropriate to describe backward induction as a refinement of the Nash equilibrium, because it already figures in John von Neumann's 'Zur Theorie der Gesellschaftsspiele', *Mathematische Annalen*, 100 (1928), 295–320 in which he proves the minimax theorem, as well as in concrete studies of chess by Ernst Zermelo, 'Über eine Anwendung der Mengenlehre auf die Theorie des Schachspiels', in E. Hobson and A. Love (eds.), *Proceedings of the Fifth International Congress of Mathematicians, Held at Cambridge 22–28 August, 1912: Volume II: Communications to Sections II–IV* (Cambridge: Cambridge University Press, 1913), 501–504, Dénes König 'Über eine Schlussweise aus dem Endlichen ins Unendliche', *Mitteilungen der Universität Szeged*, 3 (1927), 121–130, Max Euwe (the Dutch world chess champion 1935–1937), 'Mengentheoretische Betrachtungen über das Schachspiel', *Proceedings van de Koninklijke Akademie van Wetenschappen te Amsterdam*, 32 (1929), 633–642, and more fully in John von Neumann and Oscar Morgenstern, op. cit.

[55] 'Reexamination'.

[56] Ibid. 33.

equilibrium, and that, finally, perfect equilibria can be defined in an equivalent yet simpler fashion.[57]

Similar mathematics-driven model-tinkering can be found in Myerson's work on the proper equilibrium.[58] Myerson's paper first applies Selten's perfect equilibrium to normal form games, and grasps it in a mathematically simple way. He then introduces his own concept of proper equilibrium in exactly the same way that he rendered Selten's perfect equilibrium. Myerson, too, then considers sequences of games satisfying certain equilibrium conditions in terms of the ε-proper equilibrium requiring players to assign a factor of $\frac{1}{\varepsilon}$ higher probability to their 'better' choices; and he, too, defines the ultimate concept in terms of a limit, concluding with a proof of an existence theorem to the effect that every normal form game possesses at least one proper equilibrium. And conceptual and empirical motivation is absent to an even greater degree than in Selten's more elaborate 1975 paper.

5.2.3 Application-Driven Mathematisation in the Epistemic Programme

NERP may be overmathematised and introverted, but what about the Epistemic Programme itself? The systematic internal study of epistemic characterisation results has shown that the development of many modelling techniques was motivated by applications rather than by purely mathematical reasons. It is instructive to conclude with a more historical summary of the three lines of research in the Epistemic Programme that I discussed earlier.

The first line of research is represented by the work of Wolfgang Spohn, Douglas Bernheim and David Pearce.[59] Their main theoretic contribution is to have put epistemic concerns explicitly on the research agenda of the theory of games. The refinement researchers were never prompted to take on the task of formalising the beliefs of the players of games, simply because they kept their epistemic assumptions implicit. In the writings of Spohn, Bernheim and Pearce, by contrast, explicit connections are established between formal work on the epistemic set-up of game-playing situations and the psychology of interactive strategic agency. Their starting point is that the usual arguments in favour of the Nash equilibrium often make weak, possibly even implausible assumptions about how the players decide and what they

---

[57] I cannot overemphasise that I do not at all mean to imply anything about Selten's or any other game theorist's general outlook on science. I focus on the research programmes and some of their key publications, not on the researchers.

[58] Art. cit.

[59] Wolfgang Spohn, 'How to Make Sense of Game Theory', in W. Balzer, W. Spohn, and W. Stegmüller (eds.), *Studies in Contemporary Economics: Vol. 2: Philosophy of Economics* (Berlin: Springer, 1982), 239–270, B. Douglas Bernheim, 'Rationalizable Strategic Behavior', *Econometrica*, 52 (1984), 1007–1028, and David Pearce, 'Rationalizable Strategic Behavior and the Problem of Perfection', *Econometrica*, 52 (1984), 1029–1050.

believe. Objecting to the uniqueness assumption underlying NERP, Bernheim writes that

> even if some [refinement] technique always isolated unique equilibria, it would represent a psychological hypothesis... empirically relevant only if it formalized characteristics that are already universally perceived as salient.[60]

While they did not make explicit use of empirical data or thorough conceptual analysis, Bernheim and Pearce were nonetheless well aware of the fact that insights into rational agency would lead game theory away from NERP, and their theoretic efforts are a constructive step towards more adequate and explicit modelling that aspires to a higher degree of conceptual and empirical accuracy. In Bernheim's characteristically modest words:

> Analysis of strategic economic situations requires us, implicitly or explicitly, to maintain as plausible certain psychological hypotheses. The hypothesis that real economic agents universally recognize the salience of Nash equilibria may well be less accurate than, for example, the hypothesis that agents attempt to 'out-smart' or 'second-guess' each other, believing that their opponents do likewise.[61]

Bernheim and Pearce developed static models of the beliefs of the players in terms of common belief about rationality and utility. Players are fully and correctly informed about the exact utility functions of their opponents as well as of the fact that they are expected utility maximisers. There is no room for mistakes, nor is there room for revising and updating false or incomplete beliefs.

A different line of research within the Epistemic Programme would turn to the study of strategic interaction in situations in which the assumptions about the beliefs, desires, and rationality of the players are relaxed so as to incorporate the fact that, as Robert Aumann has pointed out, 'common knowledge of rationality is an ideal condition that is rarely met in practice'.[62] Some of the basics were set out at around the same time that Bernheim and Pearce published their work.[63] Most papers in this second line of research date from the early 1990s, however. Eddie Dekel and Drew Fudenberg modelled situations in which players may make small mistakes about each others' utility functions, and proved an epistemic characterisation result connecting it to the decidedly non-equilibrium Dekel–Fudenberg procedure.[64] What is important, though, is that by modelling (slightly) uninformed players Dekel and Fudenberg attempted to increase the level of realism:

> Nash equilibrium and its refinements describe situations with little or no 'strategic uncertainty', in the sense that each player *knows* and is *correct* about the beliefs of the other players regarding how the game will be played. While this will sometimes be the case, it is

[60] Art. cit. 1010.

[61] Ibid.

[62] 'Backward Induction and Common Knowledge of Rationality', *Games and Economic Behavior*, 8 (1995), 18.

[63] See, e.g., Paul Milgrom, 'An Axiomatic Characterisation of Common Knowledge', *Econometrica*, 49 (1981), 219–222.

[64] For further discussion (and qualification), see Section 2.3.

> also interesting to understand what restrictions on predicted play can be obtained when the players' strategic beliefs may be inconsistent.[65]

The Epistemic Programme would have to go even further. The models of Dekel and Fudenberg and others were still incapable of describing a relatively obvious feature of rational agency—the revision of beliefs in the light of new evidence. A third line of the Epistemic Programme turned to precisely this question. A number of researchers had shown over the years that there is a need to design models of how players revise their beliefs if they encounter opponents who do not play as expected.[66] In particular, they argued that the assumption of common belief about rationality is deeply problematic if it is to be maintained across all worlds of possible play. Players may start a game with common beliefs about rationality and utility, but as soon as someone makes a choice that contradicts these beliefs, her opponents will have to revise their beliefs.[67] Subtle distinctions between various belief revision policies made it possible to prove new, more precise epistemic characterisation results. Robert Stalnaker modelled players who not only maximise their expected utility given their actual beliefs, but also given what they would believe, were they to discover that they were mistaken about their opponents' prospective strategy choices. Stalnaker connected this notion of rationality to an earlier result showing that common belief about such rationality leads directly to the Dekel–Fudenberg procedure.

I have defended the claim that the Epistemic Programme fares better than the Nash Equilibrium Refinement Programme if you measure success internally, not by reference to explanatory or predictive power or other empirical factors, but rather by reference to the way they contribute to the game-theoretic understanding of the strategic interaction of rational players. I have also explained the different levels of success by noting that the Nash Equilibrium Refinement Programme shows overmathematisation, drawing its main inspiration from mathematical concerns rather than the application-driven concerns that inspire the Epistemic Programme. This does not mean that the Epistemic Programme affords concrete, empirical insights into economic agency. But it does underline that, as it stands, it is a flourishing area of interdisciplinary work on the conceptual study of interactive rational agency.

---

[65] Ibid. 243 (emphasis in original).

[66] See references in Chapters 2 and 3.

[67] See Sections 2.3 and 2.4.

Conclusion

In the first part of this book I considered game theory from the perspective of action theory, and I developed apparatus, made more formally precise in the second part, to give detailed epistemic characterisations of several game-theoretic solution concepts. The third part of the book, by contrast, presented somewhat pessimistic conclusions about game theory and its prospects as a social theory: nonsensical as a normative theory, reducible and epistemologically narrow as a descriptive theory, and also a home to a Nobel Prize winning research programme that was overmathematised, introverted and unacceptably dependent on model-tinkering. Why then should one care about epistemic characterisation theorems in the first place?

The simplest answer is to say that the Epistemic Programme is not covered by the criticism. What is normatively nonsensical and descriptively prejudiced are explanations of human interaction that use a game-theoretic model plus a solution concept as their two sole ingredients. Epistemic characterisation results and the like do not furnish such explanations, however. While some of the first epistemic characterisation theorems may have been directly linked to a solution concept already on the market, many investigations leave the 'solution concept–driven' approach in favour of an account that starts from epistemic conditions to investigate their behavioural consequences.

My argument was, in fact, that any truly game-theoretic explanation (one in terms of a model plus a solution concept) can be reduced to a decision-theoretic explanation which uses, as a model, exactly the epistemic conditions and the rationality principle the relevant epistemic characterisation theorem postulates. The Epistemic Programme is not left untouched by my critique. Rather, it is essential to make my philosophical argument work. As Adam Brandenburger has recently expressed it:

> A largely open area is to find logics that allow us to carry out epistemic analyses like the ones discussed... Such logics must be able to express concepts such as rationality, strong belief, assumption, etc., allow the existence of complete structures, and yield conclusions about solution concepts... An analysis of this type would have the benefit of being much more explicit about the players' reasoning processes in games.[1]

[1] 'The Power of Paradox: Some Recent Developments in Interactive Epistemology', *International Journal of Game Theory*, 35 (2007), 487.

B. de Bruin, *Explaining Games*, Synthese Library 346,

DOI 10.1007/978-1-4020-9906-9,

The first part of this book is a modest contribution to this area.

While epistemic characterisation theorems may not display the logical or action-theoretic problems I revealed for game theory as a descriptive enterprise, one may still ask whether they can be truly useful in actual social scientific theorising. Here I am probably less optimistic than representatives of the Epistemic Programme such as Adam Brandenburger. While some of the epistemic assumptions studied in the Epistemic Programme can easily be set aside because of inconsistency or incompatibility (witness my analysis of the debate on backward induction), many of those that are perfectly acceptable from a logical point of view are dubious if one thinks of them as describing real human strategic agents. For instance, when will it be the case that a set of agents' epistemic assumptions are exactly of the form postulated by the epistemic characterisation theorem of the Dekel–Fudenberg procedure? In the light of evidence from research programmes such as evolutionary game theory, behavioural game theory and stochastic game theory I advocate cooperation between these programmes. Evolutionary game theory fares better descriptively in situations of iterated game-playing and evolutionary learning, while behavioural game theory provides more adequate models of many one-shot cases, and much can be gained from combining these approaches with the Epistemic Programme.

A more principled motivation underlies the transition between the two parts, though. As I have stressed on several occasions, the main aim of the research project documented in this book was to look at non-cooperative game theory from the perspective of a philosopher of social science, and this project was subsumed in two natural projects. The first project was to investigate non-cooperative game theory internally—to study the assumptions it makes about strategic agents, about their beliefs, their rationalities, and so on. To carry out this project, much use was made of the results obtained by game theorists themselves in the Epistemic Programme. These results had to be amplified for my purposes by correcting them (Dekel–Fudenberg procedure), by developing an epistemic logical system that connects them more closely to the action-theoretic and philosophy of science questions with which I am concerned, and by comparing them with the more informal claims made in game-theoretic papers. Reconstructing Aumann's and Reny's approaches to backward induction, for instance, I suggested that both lead to inconsistent positions concerning extensive game-play. Furthermore, I had to interpret them, which in fact forms the core of the first part of the book. In other words, the outcome of the first project was a discussion of the internal workings of non-cooperative game theory facilitated by logical formalism tailored to the interpretative philosophical questions about what kind of human agents game theory really models.

The second project was to examine game theory externally—to study how it relates, or can relate, to human strategic interaction. First, logical analysis revealed that solution concept–driven game theory makes no sense as a normative theory of human agency, and that it can be reduced to decision theory as a descriptive theory. Second, I undertook a close reading of a game-theoretic literature and argued that a true-in-the-abstract view of scientific modelling was adopted by researchers in the Nash Equilibrium Refinement Programme which led them to be guided by mathematical concerns rather than by applications. A detailed analysis led to a vindication of the Epistemic Programme.

In conclusion, I will mention a number of important questions left open for future research. In the first two chapters I presented a logical analysis of the form of decision-theoretic and game-theoretic explanations, and I developed the original distinction between a one-shot and a many-moment interpretation of extensive game-play. While in Chapter 3 I formalised belief revision in the context of the many-moment interpretation, I did not give a formalisation of the many-moment interpretation as such. It is a task for future research to give a formal version of these two interpretations and their possible subdivisions, and to see whether one could think of an appropriate notion of equivalence of game-playing situations to show interrelations between particular versions of the one-shot interpretation and the many-moment interpretation.[2]

In Chapter 2 I discussed normal form game solution concepts and interpreted their epistemic and rationality assumptions. I omitted such concepts as the approachable, correlated, essential, firm, regular, and strictly and weakly proper equilibrium. I dealt only with the correlated equilibrium in Chapter 4 and with the proper equilibrium in Chapter 5. While my critical remarks about the Nash Equilibrium Refinement Programme apply, I believe, to all of these refinement concepts, it would nonetheless be an interesting project to try and extend my inductive and implicit formalisation methods to these concepts and to continue the internal investigations.

In Chapter 3 I analysed two approaches to backward induction, but I did not connect the findings to a longstanding problem in game theory, namely, the Backward Induction Paradox. I touched upon the paradox in Chapter 5 in the context of introversion and model-tinkering when I discussed Selten's paper on 'The Chain-Store Paradox' and I mentioned a solution to the paradox presented in the literature—a highly tinkered one.[3] I believe that much can be gained here by making a clear distinction between the one-shot interpretation and the many-moment interpretation of extensive game-play. If I am right, an important question is to investigate how the players perceive the game they are playing.

To get some impression of what I mean, consider what Graham Priest states in a paper on backward induction, referring to a Centipede game such as the one shown in Figure 1:

> [I]n the centipede, it is clear that we can *both* be better off if [the first player] does not terminate the game at the first move...
>
> [A]ccording to the game-theoretic orthodoxy, however, the backwards induction is correct in the centipede game and its ilk, and thus the game should finish at the first move.[4]

---

[2] On game equivalence, see Boudewijn de Bruin, 'Game Transformations and Game Equivalence', ILLC Technical Note X-1999-01 (University of Amsterdam, 1999), and Susan Elmes and Philip Reny, 'On the Strategic Equivalence of Extensive Form Games', *Journal of Economic Theory*, 62 (1994), 1–23.

[3] *Theory and Decision*, 9 (1978), 127–159.

[4] 'The Logic of Backwards Inductions', *Economics and Philosophy*, 16 (2000), 268 (emphasis in original).

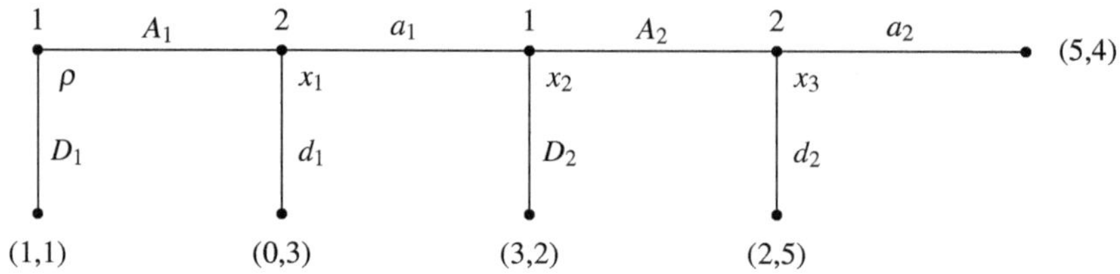

Fig. 1

The analysis of the logical form of game-theoretic explanations and the meaningless normative interpretations shows that this is misguided, though. Certainly, the two players of the Centipede would be better off at almost all of the other terminal nodes than at the first (inductive) terminal node, but the Centipede is not intended to capture the behaviour of groups of players as groups, because it is a model of individual preferences and motivations. As I have suggested, the normative interpretation as advice to the two players as a group in general is just bad advice. For individual players the situation may be more subtle, however. Of course, almost all non-inductive terminal nodes are better than the inductive one, but the peculiar structure of the game entails that the players always prefer the immediately succeeding terminal node of their decision nodes over the next succeeding terminal node. This entails that it is at least very imprecise to say that '*both* players [would be] better off if [the first player] do[es] not terminate the game at the first move', because the first player would not be better off if the game terminated in the second terminal node. In other words, the idea that backward induction arguments are intended to show that 'rational' players play inductive strategies is careless.

Appendix A
Notation, Definitions, Theorems

A.1 Decision Theory

A decision maker i has a finite set of *states* Ω and a finite set of (pure) *consequences* C. In fact, finiteness need not be assumed as long as one requires that the probability measures only give positive weight to finitely many elements from C. The set of probability measures over C is written $\Delta(C)$ and its elements are called *lotteries* over C. Agent i further has a (weak total) *preference ordering* $\succeq$ over *acts* in $\Delta(C)^{\Omega}$, the set of maps from Ω to $\Delta(C)$. If i chooses some act $f\colon \Omega \to \Delta(C)$, then if ω is the actual state of the world (that is, if nature plays ω), the outcome is determined by the lottery $f(\omega)$. The *conditional preference ordering* $\succeq_S$ for any event $S \subseteq \Omega$ is defined by

$$f \succeq_S g \text{ whenever there is an } h \text{ such that } (f_S, h_{-S}) \succeq (g_S, h_{-S}),$$

and an event S (a subset of Ω) is called *Savage-null* if and only if $f \sim_S g$ for all f and g. A *constant act* maps every state to the same lottery. I follow the convention to use $\succ$ as abbreviating the strict pendant of $\succeq$, and $\sim$ for the induced similarity relation. The following axioms are the von Neumann–Morgenstern axioms.[1]

Ord $\succeq$ is a total, reflexive and transitive ordering; that is, for all f, g and h, $f \succeq g$ or $g \succeq f$, $f \succeq f$, and $f \succeq g$ and $g \succeq h$ only if $f \succeq h$.

OI The acts are objectively independent; that is, for all f, g, h, and $0 < \alpha \leq 1$, if $f \succ g$, then $\alpha f + (1-\alpha)h \succ \alpha g + (1-\alpha)h$, and similarly for $\sim$.

NT The ordering is non-trivial; that is, there are f and g such that $f \succ g$.

Arc The ordering is Archimedean; that is, for all f, g and h, if $f \succ g \succ h$, then there exists $0 < \alpha < \beta < 1$ such that $\beta f + (1-\beta)h \succ g \succ \alpha f + (1-\alpha)h$.

NN For all $\omega_1, \omega_2 \in \Omega$ that are not Savage-null, and constant acts f and g, $f \succeq_{\{\omega_1\}} g$ if and only if $f \succeq_{\{\omega_2\}} g$.

The following representation theorem is standard.

[1] Lawrence Blume, Adam Brandenburger and Eddie Dekel, 'Lexicographic Probabilities and Choice Under Uncertainty', *Econometrica*, 59 (1991), 61–79.

Theorem A.1 (Blume et al., 1991) *The above axioms hold if and only if there are a linear function* $u\colon \Delta(C) \to \mathbb{R}$ *and a probability measure* P *on* Ω *such that for all acts* f *and* g *it is true that*

$$f \succeq g \text{ if and only if } \sum_{\omega\in\Omega} \mathsf{P}(\omega)u(f(\omega)) \geq \sum_{\omega\in\Omega} \mathsf{P}(\omega)u(g(\omega)).$$

Furthermore, u *is unique up to positive linear transformations,* P *is unique, and* $\mathsf{P}(S) = 0$ *if and only if* S *is Savage-null.*

The axiom about Archimedeanness can be weakened in order to talk about *lexicographic probability systems*. These are N-tuples of probability measures over the set of states Ω for integers N with *lexicographic ordering* $\geq_L$ of $\mathbb{R}^N$ defined as follows: $(a_1,\ldots,a_N) \geq_L (b_1,\ldots,b_N)$ if and only if whenever $b_i > a_i$ there is a $j < i$ such that $a_j > b_j$. The alternative axiom is a conditional Archimedean property.

CArc For each $\omega \in \Omega$ and f, g and h, if $f \succ_{\{\omega\}} g \succ_{\{\omega\}} h$, then there exist $0 < \alpha < \beta < 1$ such that $\beta f + (1-\beta)h \succ_{\{\omega\}} g \succ_{\{\omega\}} \alpha f + (1-\alpha)h$.

The corresponding alternative representation theorem is this.

Theorem A.2 (Blume et al., 1991) *The above axioms (with CArc instead of Arc) hold if and only if there are a linear function* $u\colon \Delta(C) \to \mathbb{R}$ *and a lexicographic probability system* $(\mathsf{P}_1,\ldots,\mathsf{P}_N)$ *on* Ω *such that for all acts* f *and* g *it is true that*

$$f \succeq g \text{ if and only if } \left(\sum_{\omega\in\Omega} \mathsf{P}_i(\omega)u(f(\omega))\right)_{i=1}^{N} \geq_L \left(\sum_{\omega\in\Omega} \mathsf{P}_i(\omega)u(g(\omega))\right)_{i=1}^{N}.$$

Furthermore, u *is unique up to positive linear transformations, the cardinality of* Ω *is an upper bound for* K*, the* P_i *are unique up to linear combinations among them, and* $\mathsf{P}_i(S) = 0$ *for all* i *if and only if* S *is Savage-null.*

A.2 Normal Form Games

A *normal form game* Γ is a tuple of the form $(I,(A_i)_i,(\succeq_i)_i)$, where I is the set of *players*, A_i the set of *actions* from which player i can and has to choose, and $\succeq_i$ the *preference relation* of player i. In general it is assumed that the set of players as well as the sets of actions are finite. The players of a game are usually represented by numerals 1, 2, 3, and so forth. The actions player i can choose between in some normal form game are written i_1, i_2, i_3, and so forth, assuming some arbitrary enumeration. The preference orderings have to satisfy the above axioms for non-lexicographic probabilities. Due to Theorem A.1 an equivalent definition in terms of *utility functions* can be given. A normal form game then is a tuple $(I,(A_i)_i,(u_i)_i)$, where the u_i are real-valued utility functions on $\prod_j A_j$.

Normal form games for two players are nicely represented by matrices, and normal form games for three players can be represented by tuples of matrices, a kind of *multi-matrix*. Although without a visually attractive counterpart, it is equally pleasant for games with more than three players to write them down by means of a multi-matrix. For that reason any N-person normal form game is written $(p_{i,k_1,\ldots,k_N})_{i,k_1,\ldots,k_N}$. These numbers are nothing more than the utility values for all possible combinations, or

$$u_i(1_{k_1},\ldots,N_{k_N}) = p_{i,k_1,\ldots,k_N}.$$

The number of players can be read off easily from the number of arguments; how many actions each player i can take can be read off (a bit more indirectly) from the range of the k_i. With more precision, the elements from A_i are i's *pure* actions. Linear combinations (probability measures) over A_i are called *mixed* actions. Because of the von Neumann–Morgenstern axioms, the utility of a mixed action profile is equal to the weighted utilities of the pure actions. That is, utility functions are linear in actions. For games with N players, *action profiles* are tuples of the form $(a_1,\ldots,a_N)$. Tuples of the form $(a_1,\ldots,a_{i-1},a_{i+1},\ldots,a_N)$ are called *i-deleted action profiles*. Given some action profile a, the obvious i-deleted action profile is written a_{-i}. The original profile is retrieved by writing (a_i,a_{-i}).

A.3 Extensive Games

An *extensive (form) game* (with *perfect information*) Γ with players from I is based on a finite tree $(X,\prec,\rho)$ where ρ is the root or starting point of the game, and $\prec$ an obvious strict (that is irreflexive and transitive) partial ordering of the nodes, called *decision nodes*, such that $\rho \prec x$ for all $x \neq \rho$. The inverse is written $\succ$. So $x \prec y$ if y is reached later in the game so to speak. Nodes x without $y \succ x$ are called *terminal* nodes. The *depth* $d(x)$ of a decision node x is the maximum number of edges connecting x with a terminal node (an inductive definition would make this more precise). A function ι associates all decision nodes D with elements from I indicating which nodes are within a player's control. With a bit more precision one would start from some class of objects (say the natural numbers). Then one would consider the class of all trees made up from these objects. One would define equivalence relations on these trees (making all trees of the same form equivalent) to say that the game trees $(X,\prec,\rho)$ are precisely the equivalence classes under this relation.[2]

Players have preference orderings over the set of terminal nodes satisfying the von Neumann–Morgenstern axioms. Alternatively, using Theorem A.1, utility functions $u_i\colon X\backslash D \to \mathbb{R}$ can be used. It is standard to allow for some ambiguity concerning the domain of u_i ranging over strategy profiles or over terminal nodes. An extensive game Γ is called *generic* whenever all u_i are injective. Sometimes it is as-

[2] Harold Kuhn, 'Extensive Games and the Problem of Information', in ibid. and A. Tucker (eds.), *Contributions to the Theory of Games: Volume II* (Princeton: Princeton University Press, 1953), 193–216.

sumed that the preference ordering be strict (that is, irreflexive and transitive). Such an ordering leads to generic games.

A *(full) strategy* of player i is a function mapping all her decision nodes to immediate successors. That is, a function $s\colon \iota^{-1}(i) \to X$ such that $x \prec s(x)$ but $x \prec y \prec s(x)$ for no y. In general I use the term *(full) strategy* for the function s on $\iota^{-1}(i)$ (a function defined on all decision nodes of player i) and restrict the use of the term *(individual) action* to the notion of the restriction of s to one single decision node of player i. As for normal form games I write i_1, i_2, i_3, and so forth, for i's strategies on the basis of some arbitrary enumeration.

The *subgame* generated by x is simply the game based on the tree generated by x with the obvious restrictions for the player function and the utility functions. If Γ is some extensive game, the subgame generated by some decision node x is written Γ_x. The *normal form* $\mathrm{nf}(\Gamma)$ of an extensive game Γ is the normal form game where the actions (in the normal form) are the full strategies (from the extensive game), and where the utility functions are defined in the obvious way.

Bibliography

Alchourrón, C., Gärdenfors, P., and Makinson, D., 'On the Logic of Theory Change: Partial Meet Functions for Contraction and Revision', *Journal of Symbolic Logic*, 50 (1985), 510–530.

Apt, Krzysztof R., 'The Many Faces of Rationalizability', *The B.E. Journal of Theoretical Economics*, 7 (2007), article 18.

Archer, M., and Tritter, J. (eds.), *Rational Choice Theory: Resisting Colonization* (London: Routledge, 2000).

Asheim, G., 'Proper Rationalizability in Lexicographic Beliefs', *International Journal of Game Theory*, 30 (2002), 452–478.

Aumann, R., 'What is Game Theory Trying to Accomplish?', in K. Arrow and S. Honkapohja (eds.), *Frontiers of Economics* (Oxford: Blackwell, 1987), 28–76.

Aumann, R., 'Correlated Equilibrium as an Expression of Bayesian Rationality', *Econometrica*, 55 (1987), 1–18.

Aumann, R., 'Backward Induction and Common Knowledge of Rationality', *Games and Economic Behavior*, 8 (1995), 6–19.

Aumann, R., 'Reply to Binmore', *Games and Economic Behavior*, 17 (1996), 138–146.

Aumann, R., 'On the Centipede Game', *Games and Economic Behavior*, 23 (1998), 97–105.

Aumann, R., 'Interactive Epistemology I' (two parts), *International Journal of Game Theory*, 28 (1999), 263–300, 301–314.

Aumann, R., and Brandenburger, A., 'Epistemic Conditions for Nash Equilibrium', *Econometrica*, 63 (1995), 1161–1180.

Aumann, R., and Drèze, J., 'When All is Said and Done, How Should You Play and What Should You Expect?', CORE Discussion Paper 2005–21 (University of Louvain, 2005).

Bacharach, M., *Beyond Individual Choice: Teams and Frames in Game Theory*, ed. N. Gold and R. Sugden (Princeton: Princeton University Press, 2006).

Basu, K., 'Strategic Irrationality in Extensive Games', *Mathematical Social Sciences*, 15 (1988), 247–260.

Battigalli, P., and Bonanno, G., 'Recent Results on Belief, Knowledge and the Epistemic Foundations of Game Theory', *Research in Economics*, 53 (1999), 149–225.

Becker, G., *The Economic Approach to Human Behavior* (Chicago: University of Chicago Press, 1976).

Ben-Porath, E., 'Rationality, Nash Equilibrium and Backwards Induction in Perfect-Information Games', *Review of Economic Studies*, 64 (1997), 23–46.

Benthem, J. van, 'Games in Dynamic-Epistemic Logic', *Bulletin of Economic Research*, 53 (2001), 219–248.

Benthem, J. van, 'Extensive Games as Process Models', *Journal of Logic, Language and Information*, 11 (2002), 289–313.

Bernheim, B., 'Rationalizable Strategic Behavior', *Econometrica*, 52 (1984), 1007–1028.

Bicchieri, C., 'Common Knowledge and Backward Induction: A Solution to the Paradox', in M. Vardi (ed.), *Proceedings of the Second Conference on Theoretical Aspects of Reasoning about Knowledge: March 7–9, 1988, Pacific Grove, California* (Los Altos, Calif.: Morgan Kaufmann, 1988), 381–393.

Bicchieri, C., 'Self-Refuting Theories of Strategic Interaction: A Paradox of Common Knowledge', *Erkenntnis*, 30 (1989), 69–85.

Bicchieri, C., 'Rationality and Game Theory', in A. Mele and P. Rawling (eds.), *The Handbook of Rationality* (Oxford: Oxford University Press, 2003), 182–205.

Binmore, K., 'Modeling Rational Players: Part I', *Economics and Philosophy*, 3 (1987), 179–214.

Binmore, K., 'A Note on Backward Induction', *Games and Economic Behavior*, 17 (1996), 135–137.

Binmore, K., 'Rationality and Backward Induction', *Journal of Economic Methodology*, 4 (1997), 23–41.

Blume, L., Brandenburger, A., and Dekel, E., 'Lexicographic Probabilities and Choice Under Uncertainty', *Econometrica*, 59 (1991), 61–79.

Board, O., 'Dynamic Interactive Epistemology', *Games and Economic Behavior*, 49 (2004), 49–80.

Börgers, T., 'Weak Dominance and Approximate Common Knowledge', *Journal of Economic Theory*, 64 (1994), 265–276.

Boudon, R., 'La rationalité axiologique: Une notion essentielle pour l'analyse des phénomènes normatifs', *Sociologie et societés*, 31 (1999), 103–117.

Brandenburger, A., 'Knowledge and Equilibrium in Games', *Journal of Economic Perspectives*, 6 (1992), 83–101.

Brandenburger, A., 'Lexicographic Probabilities and Iterated Admissibility', in P. Dasgupta, D. Gale, O. Hart and E. Maskin (eds.), *Economic Analysis of Markets and Games: Essays in Honor of Frank Hahn* (Cambridge, Mass.: MIT Press, 1992), 282–276.

Brandenburger, A., 'The Power of Paradox: Some Recent Developments in Interactive Epistemology', *International Journal of Game Theory*, 35 (2007), 465–492.

Brandenburger, A., and Dekel, E., 'Rationalizability and Correlated Equilibria', *Econometrica*, 55 (1987), 1391–1402.

Brandenburger, A., and Friedenberg, A., 'Intrinsic Correlation in Games', *Journal of Economic Theory*, 141 (2008), 28–67.

Brandenburger, A., Friedenberg, A., and Keisler, H. J., 'Admissibility in Games', *Econometrica*, 76 (2008), 307–352.

Brandts, J., and Holt, C., 'Adjustment Patterns and Equilibrium Selection in Experimental Signalling Games', *International Journal of Game Theory*, 22 (1993), 279–302.

Broome, J., 'Normative Requirements', *Ratio*, 12 (1999), 398–419.

Broome, J., and Rabinowicz, W., 'Backwards Induction in the Centipede Game', *Analysis*, 59 (1999), 237–242.

Bruin, B. de, 'Game Transformations and Game Equivalence', ILLC Technical Note X-1999-01 (University of Amsterdam, 1999).

Bruin, B. de, 'Explaining Games: On the Logic of Game Theoretic Explanations', Diss. (University of Amsterdam, 2004).

Bruin, B. de, 'Game Theory in Philosophy', *Topoi*, 24 (2005), 197–208.

Bruin, B. de, 'Popper's Conception of the Rationality Principle in the Social Sciences', in I. Jarvie, K. Milford and D. Miller (eds.), *Karl Popper: A Centenary Assessment: Selected Papers from Karl Popper 2002: Volume III: Science* (Aldershot: Ashgate, 2006), 207–215.

Bruin, B. de, 'Common Knowledge of Payoff Uncertainty in Games', *Synthese*, 163 (2008), 79–97.

Bruin, B. de, 'Common Knowledge of Rationality in Extensive Games', *Notre Dame Journal of Formal Logic*, 49 (2008), 261–280.

Bruin, B. de, 'Epistemic Logic and Epistemology', in V. Hendricks and D. Pritchard (eds.), *New Waves in Epistemology* (Basingstoke: Palgrave Macmillan, 2008), 106–136.

Bruin, B. de, 'Reducible and Nonsensical Uses of Game Theory', *Philosophy of the Social Sciences*, 38 (2008), 247–266.

Bruin, B. de, 'On the Narrow Epistemology of Game Theoretic Agents', in O. Majer, A.-V. Pietarinen and T. Tulenheimo (eds.), *Games: Unifying Logic, Language, and Philosophy* (Dordrecht: Springer, 2009), 27–36.
Bruin, B. de, 'Overmathematisation in Game Theory: Pitting the Nash Equilibrium Refinement Programme against the Epistemic Programme', *Studies in History and Philosophy of Science Part A*, 40 (2009), 290–300.
Butters, G., 'Equilibrium Distributions of Sales and Advertising Prices', *Review of Economic Studies*, 44 (1977), 465–491.
Camerer, C., *Behavioral Game Theory: Experiments in Strategic Interaction* (Princeton: Princeton University Press, 2003).
Chisholm, R., 'Contrary-to-Duty Imperatives and Deontic Logic', *Analysis*, 24 (1963), 33–36.
Clausing, T., 'Doxastic Conditions for Backward Induction', *Theory and Decision*, 54 (2003), 315–336.
Clausing, T., 'Belief Revision in Games of Perfect Information', *Economics and Philosophy*, 20 (2004), 89–115.
Colman, A., 'Cooperation, Psychological Game Theory, and Limitations of Rationality in Social Interaction', *Behavioral and Brain Sciences*, 26 (2003), 139–198.
Conitzer, V., and Sandholm, T., 'New Complexity Results about Nash Equilibria', *Games and Economic Behavior*, 63 (2008), 621–641.
Cozic, M., 'Impossible States at Work: Logical Omniscience and Rational Choice', in R. Topol and B. Walliser (eds.), *Cognitive Economics: New Trends* (Amsterdam: Elsevier, 2007), 47–68.
Damme, E. van, *Stability and Perfection of Nash Equilibria: Second, Revised and Enlarged Edition* (Berlin: Springer, 1987).
Davis, D., and Holt, C., *Experimental Economics* (Princeton: Princeton University Press, 1993).
Davis, M., *Game Theory: A Nontechnical Introduction* (1970; rev. ed. 1983; repr. 1997, Mineola: Dover).
Dekel, E., and Fudenberg, D., 'Rational Behavior with Payoff Uncertainty', *Journal of Economic Theory*, 70 (1990), 243–267.
Dimand, M. A., and Dimand, R., *A History of Game Theory: Volume I: From the Beginnings to 1945* (London: Routledge, 1996).
Dufwenberg, M., Norde, H., Reijnierse, H., and Tijs, S., 'The Consistency Principle for Set-Valued Solutions and a New Direction for Prescriptive Game Theory', *Mathematical Methods of Operations Research*, 54 (2001), 119–131.
Elmes, S., and Reny, P., 'On the Strategic Equivalence of Extensive Form Games', *Journal of Economic Theory*, 62 (1994), 1–23.
Ernst, Z., 'Explaining the Social Contract', *British Journal for the Philosophy of Science*, 52 (2001), 1–24.
Euwe, M., 'Mengentheoretische Betrachtungen über das Schachspiel', *Proceedings van de Koninklijke Akademie van Wetenschappen te Amsterdam*, 32 (1929), 633–642.
Fagin, R., and Halpern, J., 'Reasoning About Knowledge and Probability', *Journal of the Association for Computing Machinery*, 41 (1994), 340–367.
Fagin, R., and Halpern, J., Moses, Y., and Vardi, M., *Reasoning about Knowledge* (Cambridge, Mass.: MIT Press, 1995).
Friedell, M., 'On the Structure of Shared Awareness', *Behavioral Science*, 14 (1969), 28–39.
Friedman, M., 'The Methodology of Positive Economics', in M. Friedman (ed.), *Essays in Positive Economics* (Chicago: University of Chicago Press, 1953), 3–43.
Fudenberg, D., and Tirole, J., *Game Theory* (Cambridge, Mass.: MIT Press, 1991).
Giocoli, N., *Modeling Rational Agents: From Interwar Economics to Early Modern Game Theory* (Cheltenham: Edward Elgar, 2003).
Goeree, J., and Holt, C., 'Stochastic Game Theory: For Playing Games, Not Just for Doing Theory', *Proceedings of the National Academy of Sciences*, 96 (1999), 10564–10567.
Guala, F., 'The Logic of Normative Falsification: Rationality and Experiments in Decision Theory', *Journal of Economic Methodology*, 7, 59–93.
Guala, F., 'Building Economic Machines: The FCC Auctions', *Studies in History and Philosophy of Science Part A*, 32 (2001), 453–477.

Guala, F., 'Has Game Theory been Refuted?', *Journal of Philosophy*, 103 (2006), 239–263.

Gul, F., 'Rationality and Coherent Theories of Strategic Behavior', *Journal of Economic Theory*, 70 (1996), 1–31.

Hait, W., August, D., and Haffty, B. (eds.), *Expert Consultations in Breast Cancer: Critical Pathways and Clinical Decision Making* (New York: Marcel Dekker, 1999).

Halpern, J., 'The Relationship between Knowledge, Belief, and Certainty', *Annals of Mathematics and Artificial Intelligence*, 4 (1991), 301–322.

Halpern, J., 'Substantive Rationality and Backward Induction', *Games and Economic Behavior*, 37 (2001), 321–339.

Harsanyi, J., 'Games with Incomplete Information Played by "Bayesian" Players', *Management Science*, 14 (1967–1968), 159–182, 320–334, 486–502.

Hausman, D., 'Revealed Preference, Belief, and Game Theory', *Economics and Philosophy*, 16 (2000), 99–115.

Hausman, D., 'Testing Game Theory', *Journal of Economic Methodology*, 12 (2005), 211–223.

Hees, M. van, 'Liberalism, Efficiency, and Stability: Some Possibility Results', *Journal of Economic Theory*, 88 (1999), 294–309.

Heifetz, A., and Mongin, P., 'Probability Logic for Type Spaces', *Games and Economic Behavior*, 25 (2001), 31–53.

Herings, P. J.-J., and Vannetelbosch, V., 'The Equivalence of the Dekel–Fudenberg Iterative Procedure and Weakly Perfect Rationalizability', *Economic Theory*, 15 (2000), 677–687.

Hintikka, J., *The Principles of Mathematics Revisited* (Cambridge: Cambridge University Press, 1996).

Hollis, M., *The Philosophy of Social Science: An Introduction* (Cambridge: Cambridge University Press, 1994).

Holt, C., and Roth, A., 'The Nash Equilibrium: A Perspective', *Proceedings of the National Academy of Sciences*, 101 (2004), 3999–4002.

Israel, G., 'The Science of Complexity: Epistemological Problems and Perspectives', *Science in Context*, 18 (2005), 479–509.

Jacobsen, H. J., 'On the Foundations of Nash Equilibrium', *Economics and Philosophy*, 12 (1996), 67–88.

Jiborn, M., and Rabinowicz, W., 'Backward Induction without Full Trust in Rationality', in W. Rabinowicz (ed.), *Value and Choice: Some Common Themes in Decision Theory and Moral Philosophy: Volume 2* (Lund: Lund Philosophy Reports, 2001), 101–120.

Kadane, J., and Larkey, P., 'Subjective Probability and the Theory of Games', *Management Science*, 28 (1982), 113–120.

Kincaid, H., 'Formal Rationality and its Pernicious Effects on the Social Sciences', *Philosophy of the Social Sciences*, 30 (2000), 67–88.

König, D., 'Über eine Schlussweise aus dem Endlichen ins Unendliche', *Mitteilungen der Universität Szeged*, 3 (1927), 121–130.

Kreps, D., Milgrom, P., Roberts, J., and Wilson, R., 'Rational Cooperation in the Finitely Repeated Prisoners' Dilemma', *Journal of Economic Theory*, 27 (1982), 245–252.

Kreps, D., and Wilson, R., 'Reputation and Imperfect Information', *Journal of Economic Theory*, 27 (1982), 253–279.

Kreps, D., and Milgrom, P., 'Sequential Equilibria', *Econometrica*, 50 (1982), 863–894.

Kuhn, H., 'Extensive Games and the Problem of Information', in H. Kuhn and A. Tucker (eds.), *Contributions to the Theory of Games: Volume II* (Princeton: Princeton University Press, 1953), 193–216.

Kuhn, S., 'Reflections on Ethics and Game Theory', *Synthese*, 141 (2004), 1–44.

Lewis, D., *Convention* (Cambridge, Mass.: Harvard University Press, 1969).

Luce, R. D., and Raiffa, H., *Games and Decisions: Introduction and Critical Survey* (New York: Wiley, 1957).

Milgrom, P., 'An Axiomatic Characterisation of Common Knowledge', *Econometrica*, 49 (1981), 219–222.

Milgrom, P., and Roberts, J., 'Predation, Reputation, and Entry Deterrence', *Journal of Economic Theory*, 27 (1982), 280–312.

Mill, J. S., 'On the Definition of Political Economy; and on the Method of Philosophical Investigation in that Science', *London and Westminster Review*, 26 (1836), 1–29; citations are from 1844 ed. repr. in id., *Collected Works of John Stuart Mill* ed. J. Robson (Toronto: University of Toronto Press, 1967), 309–339.

Mirowski, P., 'When Games Grow Deadly Serious: The Military Influence on the Evolution of Game Theory' in C. Goodwin (ed.), *Economics and National Security: A History of their Interaction* (ann. supp. to *History of Political Economy*) (Durham, NC: Duke University Press, 1991), 227–256

Mongin, P., 'L'optimisation est-elle un critère de rationalité individuelle?', *Dialogue*, 33 (1994), 191–222.

Mongin, P., 'Le principe de rationalité et l'unité des sciences sociales', *Revue Économique*, 53 (2002), 301–323.

Moulin, H., *Game Theory for the Social Sciences* (New York: New York University Press, 1986).

Moser, P. (ed.), *Rationality in Action: Contemporary Approaches* (Cambridge: Cambridge University Press, 1996).

Myerson, R., 'Refinements of the Nash Equilibrium Concept', *International Journal of Game Theory*, 7 (1978), 73–80.

Myerson, R., *Game Theory: Analysis of Conflict* (Cambridge, Mass.: Harvard University Press, 1991).

Nasar, S., *A Beautiful Mind: A Biography of John Forbes Nash, Jr., Winner of the Nobel Prize in Economics, 1994* (New York: Simon and Schuster, 1998).

Nash, J., 'Equilibrium Points in *n*-person Games', *Proceedings of the National Academy of Sciences*, 36 (1950), 48–49.

Nash, J., 'Non-Cooperative Games', Ph.D. diss. (Princeton University, 1950).

Nash, J., 'Non-Cooperative Games' *Annals of Mathematics*, 54 (1951), 286–295.

Neumann, J. von, 'Zur Theorie der Gesellschaftsspiele', *Mathematische Annalen*, 100 (1928), 295–320.

Neumann, J. von, and Morgenstern, O., *Theory of Games and Economic Behavior* (Princeton: Princeton University Press, 1944).

Neyman, A., 'Bounded Complexity Justifies Cooperation in the Finitely Repeated Prisoners' Dilemma', *Economics Letters*, 19 (1985), 227–229.

Osborne, M., *An Introduction to Game Theory* (New York: Oxford University Press, 2003).

Osborne, M., and Rubinstein, A., *A Course in Game Theory* (Cambridge: MIT Press, 1994).

Pearce, D., 'Rationalizable Strategic Behavior and the Problem of Perfection', *Econometrica*, 52 (1984), 1029–1050.

Pettit, P., and Sugden, R., 'The Backward Induction Paradox', *The Journal of Philosophy*, 86 (1989), 169–182,

Pietarinen, A.-V., 'Games as Formal Tools versus Games as Explanations in Logic and Science', *Foundations of Science*, 8 (2003), 317–364.

Popper, K., *Logik der Forschung* (Vienna: Julius Springer, 1935).

Popper, K., 'The Myth of the Framework', in M. Notturno (ed.), *Models, Instruments and Truth* (London: Routledge, 1998), 154–184.

Priest, G., 'The Logic of Backwards Inductions', *Economics and Philosophy*, 16 (2000), 267–285.

Rabinowicz, W., 'To Have One's Cake and Eat It, Too: Sequential Choice and Expected-Utility Violations', *Journal of Philosophy*, 92 (1995), 586–620.

Rabinowicz, W., 'Grappling with the Centipede: Defence of Backward Induction for BI-Terminating Games', *Economics and Philosophy*, 14 (1998), 95–126.

Rasmusen, E., *Games and Information: An Introduction to Game Theory* (Malden: Blackwell, 1989; 4th ed. 2006)

Reny, P., 'Two Papers on the Theory of Strategic Behaviour', Ph.D. diss. (Princeton University, 1988).

Reny, P., 'Common Knowledge and Games with Perfect Information', in A. Fine and J. Leplin (eds.), *Proceedings of the Philosophy of Science Association: Volume 2* (East Lansing, Mich.: Philosophy of Science Association, 1989), 363–369.

Reny, P., 'Common Belief and the Theory of Games with Perfect Information', *Journal of Economic Theory*, 59 (1993), 257–274.

Reny, P., 'Rational Behaviour in Extensive-Form Games', *Canadian Journal of Economics*, 28 (1995), 1–16.

Risse, M., 'What is Rational about Nash Equilibria?', *Synthese*, 124 (2000), 361–384.

Rubinstein, A., 'Comments on the Interpretation of Game Theory', *Econometrica*, 59 (1991), 909–924.

Rubinstein, A., 'A Subjective Perspective on the Interpretation of Economic Theory', in A. Heertje (ed.), *The Makers of Modern Economics: Volume 1* (New York: Harvester Wheatsheaf, 1993), 67–83.

Rubinstein, A., 'Joseph Schumpeter Lecture: A Theorist's View of Experiments', *European Economic Review*, 45 (2001), 615–628.

Samuelson, L., 'Dominated Strategies and Common Knowledge', *Games and Economic Behavior*, 4 (1992), 284–313.

Samuelson, L., *Evolutionary Games and Equilibrium Selection* (Cambridge, Mass.: MIT Press, 1997).

Samuelson, P., 'A Note on the Pure Theory of Consumers' Behaviour', *Economica*, 5 (1938), 61–71.

Savage, L., *The Foundations of Statistics* (New York: Wiley, 1954).

Schwemmer, O., 'Aspekte der Handlungsrationalität: Überlegungen zur historischen und dialogischen Struktur unseres Handelns', in H. Schnädelbach (ed.), *Rationalität: Philosophische Beiträge* (Frankfurt am Main: Suhrkamp, 1984), 175–197.

Selten, R., 'Spieltheoretische Behandlung eines Oligopolmodells mit Nachfrageträgheit: Teil I: Bestimmung des dynamischen Preisgleichgewichts', *Zeitschrift für die gesamte Staatswissenschaft*, 121 (1965), 310–324.

Selten, R., 'Reexamination of the Perfectness Concept for Equilibrium Points in Extensive Games', *International Journal of Game Theory*, 4 (1975), 25–55.

Selten, R., 'The Chain-Store Paradox', *Theory and Decision*, 9 (1978), 127–159.

Selten, R., 'Comment [on R. Aumann's 'What is Game Theory Trying to Accomplish?']', in K. Arrow and S. Hohkapohja (eds.), *Frontiers of Economics* (Oxford: Blackwell, 1987), 77–87.

Sensat, J., 'Game Theory and Rational Decision', *Erkenntnis*, 47 (1998), 379–410.

Shapiro, I., *The Flight from Reality in the Human Sciences* (Princeton: Princeton University Press, 2005).

Simon, H., *Reason in Human Affair* (Stanford: Stanford University Press, 1983).

Sobel, J. 'Backward-Induction Arguments: A Paradox Regained', *Philosophy of Science*, 60 (1993), 114–133.

Sorensen, R., 'Paradoxes of Rationality', in A. Mele (ed.), *The Handbook of Rationality* (Oxford: Oxford University Press, 2004).

Spohn, W., 'How to Make Sense of Game Theory', in W. Balzer, W. Spohn and W. Stegmüller (eds.), *Studies in Contemporary Economics: Vol. 2: Philosophy of Economics* (Berlin: Springer, 1982), 239–270.

Spohn, W., 'Wie kann die Theorie der Rationalität normativ und empirisch zugleich sein', in L. Eckensberger and U. Gähde (eds.), *Ethik und Empirie: Zum Zusammenspiel von begrifflicher Analyse und erfahrungswissenschaftlicher Forschung in der Ethik* (Frankfurt am Main: Suhrkamp, 1993), 151–196.

Spohn, W., 'The Many Facets of Rationality', *Croatian Journal of Philosophy*, 2 (2002), 249–264.

Stalnaker, R., 'On the Evaluation of Solution Concepts', *Theory and Decision*, 37 (1994), 49–73.

Stalnaker, R., 'Knowledge, Belief and Counterfactual Reasoning in Games', *Economics and Philosophy*, 12 (1996), 133–163 (repr. with proofs in C. Bicchieri, R. Jeffrey and B. Skyrms (eds.), *The Logic of Strategy* (New York: Oxford University Press, 1999), 3–38).

Stalnaker, R., 'Response to Bonanno and Nehring', *Theory and Decision*, 45 (1998), 297–299.

Stalnaker, R., 'Belief Revision in Games: Forward and Backward Induction', *Mathematical Social Sciences*, 36 (1998), 31–56.

Stalnaker, R., 'Extensive and Strategic Forms: Games and Models for Games', *Research in Economics*, 53 (1999), 293–319.

Tan, T., and Werlang, S., 'The Bayesian Foundations of Solution Concepts of Games', *Journal of Economic Theory*, 45 (1988), 370–391.

Taylor, M., *Rationality and the Ideology of Disconnectedness* (Cambridge: Cambridge University Press, 2006).

Vanderschraaf, P., 'Convention as Correlated Equilibrium,' *Erkenntnis*, 42 (1995), 65–87.

Weber, M., *Wirtschaft und Gesellschaft*, ed. Marianne Weber (Tübingen: Mohr, 1921; citations are from 5th ed. 1990).

Weintraub, E. R. (ed.), *Towards a History of Game Theory* (ann. supp. to *History of Political Economy*) (Durham, NC: Duke University Press, 1992).

Weintraub, E. R., and Mirowski, P., 'The Pure and the Applied: Bourbakism Comes to Mathematical Economics', *Science in Context*, 7 (1994), 245–272.

Williamson, T., *Knowledge and its Limits* (Oxford: Oxford University Press, 2000).

Zermelo, E., 'Über eine Anwendung der Mengenlehre auf die Theorie des Schachspiels', in E. Hobson and A. Love (eds.), *Proceedings of the Fifth International Congress of Mathematicians, Held at Cambridge 22–28 August, 1912: Volume II: Communications to Sections II–IV* (Cambridge: Cambridge University Press, 1913), 501–504.

Index

MIX
Papier aus verantwortungsvollen Quellen
Paper from responsible sources
FSC® C105338

If you have any concerns about our products,
you can contact us on
ProductSafety@springernature.com

In case Publisher is established outside the EU,
the EU authorized representative is:
Springer Nature Customer Service Center GmbH
Europaplatz 3, 69115 Heidelberg, Germany

Printed by Libri Plureos GmbH
in Hamburg, Germany